transcriptions by Jerry Goebel

Second Printing

Editor: Susan Betts

Publisher and Design
Wayne Dementi
Dementi Milestone Publishing, Inc.
Manakin-Sabot, VA 23103
www.dementimilestonepublishing.com

The Library of Congress Control Number: 2015939512

ISBN: 978-0-9909613-5-2

Cover design by Sarah Lapallo Beck
Inkwell Book Co., Richmond, VA

Printed in the United States

for George

A member of the 140th N.Y. commented that "soldierly duty was his inward watchword" of his colonel, George Ryan, and Ryan's writings are positive proof. Penned during a dozen-year stretch of uninterrupted military life – from West Point and service in pre-war western territories, to his Civil War tenure as a staff officer and then commander of the 140th – they reveal a young officer with an understanding of a military life. With patient optimism and a work ethic driven by modesty, Ryan's letters illustrate his high attention to detail and provide a professional soldier's view of army service.

Brian A. Bennett, author of "Sons of Old Monroe: A Regimental History of Patrick O'Rorke's 140th N.Y.V.I."

George Ryan

Escaping from the fire
n the woods – "Wilderness"

All of the letters and documents in this book are part of the Ryan Family Papers held at The Catholic University of America. The collection contains the family's correspondence from 1827-1943. Most of the letters date from 1843 through the Civil War.

John, Mathew, Charles and Kyran Ryan sailed into the Port of New York from Kilkenny, Ireland on July 10, 1827. At the start of their American odyssey, the brothers worked in the textile mills of Massachusetts. In 1836, the Ryans purchased a woolen mill in Norfolk, Connecticut. The mill sat at the base of the Buttermilk Falls, and the water power there brought the family a burst of early prosperity.

John Ryan married Johanna Boomer Reed in 1835, and their son George was born one year later. Over the years, the family grew to include 12 siblings for George. The letters chosen for this book give a simple narrative of George Ryan's adult life, as written by the young man himself and the family he loved.

The Ryan Papers, all 1200 letters and documents, were transcribed and donated to The Catholic University by Ryan descendant Jerry Goebel in 2002. To bring an ease of reading to this work, modern punctuation has been overlaid as needed.

The Players

John Ryan is George's father

Johanna Ryan is George's mother

Charles M. "Chas" Ryan is George's uncle

Mathew (Mat) Ryan is George's uncle

George Ryan was born on April 19, 1836 at Medway, Mass.

Mathew J. (Mat) Ryan is George's brother

Frank Ryan is George's brother

John Ryan is George's brother

Bernard O'Reilly is the Catholic Bishop of Hartford

Origen S. Seymour is the U. S. Congressman (D) for Litchfield, Ct.

J & E. E. Ryan & Co. is the woolen mill owned and operated by the first-generation Ryan brothers, John, Mathew, Charles and a family friend, Edward E. Ryan

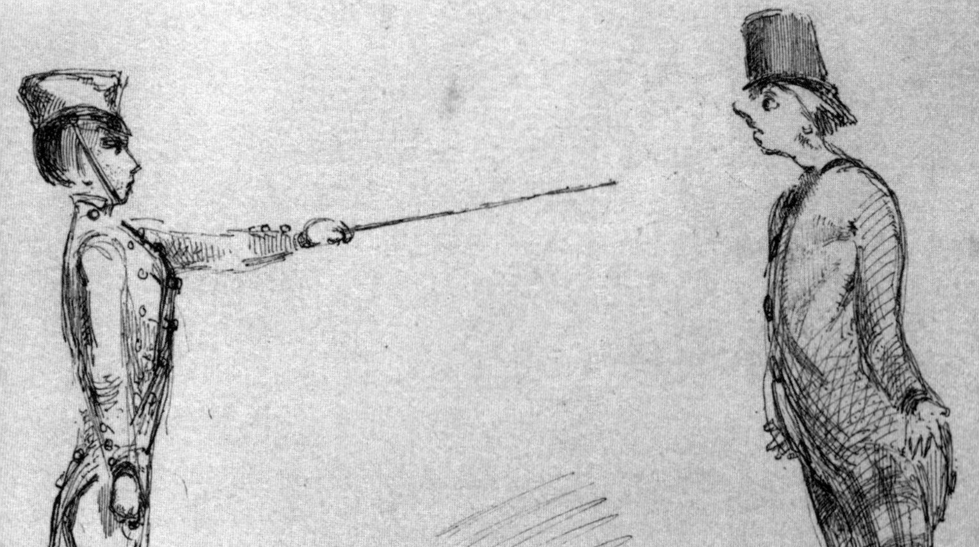

Position of a Soldier: annihilation of the Bowels.

J McN

I

West Point

Providence, October 5th 1852

Dear Sir

I write for the purpose of asking yourself, and respected partners to endeavour to have a church erected on the site selected, in the coming Spring. you might, this winter, as you intimated before to me, get out all the framing timbers, which would diminish much the expense of the building. I only intimate this to you, as, of course, you know best the proper and most feasible course, in such a matter as this.

I am most anxious, whatever way it may be accomplished, to see a church at Norfolk. religion will suffer there until then, and full justice will not be done until then, to the young generation. I expected to have seen this good work accomplished there, this Spring, but I learned with much sorrow your monetary percuniary difficulty and [unreadable] silent. I rejoice to know, that this seasonal trial has passed over, and hope that your efforts will be blessed with continuous prosperity.

GR

embark then on this good work, out of love for God, and God may give his blessing to all your undertakings.

I would like to receive the title to the Burial lot, also to the church lot, as early as convenient.

you sent me these, but as they were to myself and successors, I thought it better to have them changed, as I have no legal successors; a Corporation only has such, and the property at my my death, would go to the heirs or the grantors, or their heirs so willed. A deed in trust, in my case, would be nothing better than a life interest. I explain, that you may understand me, I wish to do all things right.

I will leave for Europe on the 16th, and pray that God may bless you during my absence. I wrote Rev[d] Mr. Smith, that I have thus directed you in the matter of the church, you may, if you wish, show him this. Respects to all the families

Most respectfully yours

+Bernard, Bp of Hartford

GR

Norfolk, January 28th, 1853

S.C. Scoville Esq.

Dear Sir:

As the time draws near when the mortgage must be put on record, if it is not withdrawn, and as we have not heard from you, or seen you here, we thought it would be no harm to write you in relation to the matter. The putting of the deed on record, will not do you any good, and it may do us a great deal of harm. In the first place, there will be, in a few days from now Forty seven thousand dollars, of the notes paid. In the second place, if not one dollar, out of the Fifty thousand had been paid the mortgage which you hold would not secure you. In point of law the instrument is not worth one cent to secure you from loss, and Mr Peck told me so when the writer went to see him about it. The writer knew this when he saw you at the Bank on Monday, but he thought that he would not mention the fact to you. Since then, he showed a true copy of the mortgage to another lawyer and he confirmed Mr Peck's opinion.

On the correctness of what we now state you may rely. We cannot for a moment suppose that you wish to injure us, in our credit to the amount of Fifty thousand dollars and that too without helping yourself to one cent but this you do by having this mortgage recorded. On our part we feel as if we carried out, in good faith, all the conditions, under which we gave the claim to you, and after doing so, we think that it is too hard to suffer in credit to such a large amount as that mortgage, put on record in its present form, would convey to the public. If you permit the deed to be recorded, with the knowledge of the facts, herein stated in your possession, we shall ever feel as if you have been quilty of terrible injustice to us. Indeed no future explanation will satisfy us. We are, Dear Sir

Very Truly Yours
J & E. E. Ryan & Co

(letter addressed to S.C. Scoville, Salisbury)

Washington Feb 11th '53

Dear Sir

I duly received yours of the 9th and will do what I can to prevent any premature change of the Postmaster in your village but can hardly believe that any man can be so shortsighted as to attempt the trick you speak of. An appointment made by the present Administration would not be a very high recommendation of the person appointed. If however the Whigs should make it a point to effect a change my influence with the present P. M. General would be very slight as you can readily see in comparison with the influence of Mr Smith and we must look mainly to the incoming Administration to correct any abuses which may be committed by the persons now in power.

The reply to your inquiry respecting a Cadet for West Point an appointment is to be made during the present month from the (4th) Fourth Congressional District A large number of candidates have been presented and I had pretty much made up my mind upon the person to be nominated but if you have any candidate to present I will endeavor to give a fair consideration to his claims. But the time is now very short and it will be necessary that the name should be presented without delay. No recommendations are necessary but a statement of the precise age of the applicant- his residence- his health- habits- education- and whether he has means of his own or of friends to get an education otherwise than at the Public expense.

Yours Very Respectfully
O. S. Seymour

To John Ryan Esq
Norfolk

Washington Feb 17th '53

Dear Sir

Yours of the 15th was duly received but I have had no application from Mr Ryan on the subject you speak of and drop you this line by return mail thinking that the application may perhaps have miscarried. Only a few days now remain before I shall be called on to make the nomination and I am desirous of giving Mr Ryan a full opportunity to present his case. I send you herewith the regulations respecting the candidates which you will please communicate without delay to Mr Ryan and let him see that they are complied with and especially the precise age of his son. I cannot give him much encouragement for receiving the appointment but would be very glad to give it to him if I properly could and circumstances may perhaps occur to enable me to do it.

Yours Very Respectfully
O. S. Seymour

To N. B. Stevens Esq
Norfolk

Washington Feb 23. 1853

My Dear Sir

I duly rec.^{d} yours of the 21st your former letter also came to hand a day or two after its due time. - There are several candidates of equal merit, I have never seen any of them. None of them are relatives of my own. I am at great loss what to do, I must decide in a few days & will communicate the result. I am anxious to oblige you & gratify your son. I shall reexamine all my papers and act as my best judgment dictates. I regret that whatever the decision is some young aspirants for education must be disappointed.

With sincere respect

O. S. Seymour

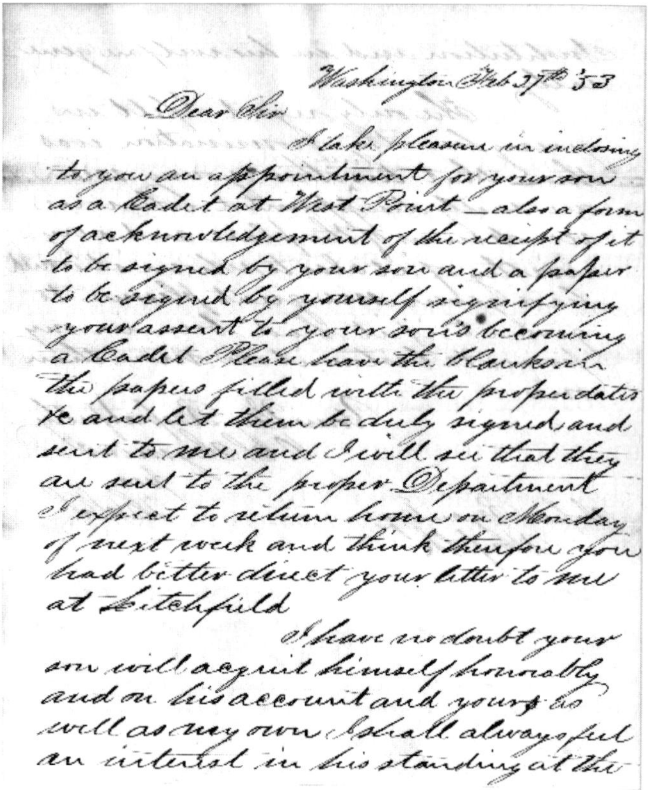
Washington Feb 27th '53

Dear Sir

I take pleasure in inclosing to you an appointment for your son as a Cadet at West Point — also a form of acknowledgement of the receipt of it to be signed by your son and a paper to be signed by yourself signifying your assent to your son's becoming a Cadet Please leave the blanks in the papers filled with the proper dates &c and let them be duly signed and sent to me and I will see that they are sent to the proper Department I expect to return home on Monday of next week and think therefore you had better direct your letter to me at Litchfield

I have no doubt your son will acquit himself honorably and on his account and yours as well as my own I shall always feel an interest in his standing at the

This 1853 letter from Congressman O. S. Seymour must have filled the Ryan household with joy!

Washington Feb 27th '53

Dear Sir

I take pleasure in inclosing to you an appointment for your son as a Cadet at West Point - also a form of acknowledgement of the receipt of it to be signed by your son and a paper to be signed by yourself signifying your assent to your son's becoming a Cadet. Please have the blanks in the papers filled with the proper dates &c and let them be duly signed and sent to me and I will see that they are sent to the proper Department. I expect to return home on Monday of next week and think therefore you had better direct your letter to me at Litchfield.

I have no doubt your son will acquit himself honorably and on his account and yours as well as my own. I shall always feel an interest in his standing at the Institution and in his welfare generally.

The only regret I felt in giving him the nomination was this- that three other young men are sadly disappointed who were very desirous of the place and very worthy of it also to whom it would have given me great pleasure to give it if it had been in my power to bestow it on more than one.

Yours Very Respectfully
O. S. Seymour

To John Ryan Esq
Norfolk

West Point N. Y. March 21st 1853

Dear Sir

In reply to your letter of the 17th I send you a printed circular which will give you most of the information desired.

You will see by it that the requirements for admission to the academy are slight. The examination in arithmetic will be strict on the subjects of vulgar and decimal fractions and Rule of Three, the candidate being required to solve any examples readily- such examples being given out by the examiner not taken from the book.

You will see also that the test for spelling is somewhat different from that ordinarily made. It would be well, if your son has not been accustomed to write from the dictation of others, to let him practice this until he finds that he can do it without making mistakes.

Though Davies' University arithmetic is used for instruction here, yet a thorough knowledge of arithmetic from any other work is as well.

GR

I would advise you to let your son come on as soon after the commencement of the month of June as convenient as he will thus, previous to his examination for admission, have an opportunity to receive instruction in the branches in which he is to be examined and to get somewhat accustomed to the ways of proceeding here.

Very Respectfully
Your Obt Servt.
A. E. Church

To John Ryan Esq
Norfolk

P. S. The examination of the Candidates takes place immediately after the examination of the classes of Cadets which will be finished about the 20th of June. may be a day or two before or after.

Village of Norfolk, Ct. circa 1890 (photo courtesy of the Norfolk Historical Society)

Bishop O'Reilly
Providence
1852

Providence, Octbr 5th 1852

Dear Sir,

I write for the purpose of advising yourself, and repeated patrons to endeavour to have a church erected on the site selected, in the coming Spring. You might this winter, as you intimated before to me, get out all the framing timbers, which would diminish much the expense of the building. I only intimate this to you, as of course you know best the proper and most feasible course in such a matter as this.

I am most anxious, whatever way it may be accomplished, to see a church at Norfolk. Religion will suffer there until then, and full justice will not

I expected to have seen this good work
accomplished here, this Spring, but
I learned with much sorrow your mom-
-entary pecuniary difficulty and dis-
-tress. I rejoice to know that this season
of trial has passed over and hope
that your efforts will be blessed with
continuous prosperity.
Embark then on this good work,
out of love for God, and God may
give his blessing to all your
undertakings.
I would like to receive the title to the
Board lot also to the church lot, as
early as convenient.
You sent me these, but as they were to
myself and successors, I thought it better
to have them changed, as I have no
legal successors; a Corporation only
has such, and the property of my

my death, would go to the heirs or
the grantors, or their heirs be called:
a deed in trust, in my case, would
be ~~worse than~~ nothing, better than a
life interest. I explain, that you
may understand me, I wish to do
all things right.
I will leave for Europe on the 16th;
and pray that God may bless you
during my absence.
I wish Revd Mr. Lynch, that I have thus
directed you in the matter of the
church, you may, if you will, show
him this: respects to all the family.
Most respectfully yours &
+ Bernard, Bp of Hartford

Dear Sir

I have to day Rec^{d} Your Letter of the 18th & it gives me pleasure to Comply with Your Request & Enclose You a Letter to my brother at West Point, who I am Shre will take pleasure in advising Your son of The best place & course to take during his stay at The Military Academy-

I hope Your Son is aware of The difficulties of getting Through The Academy & The strict disipline observed there. I trust he will apply himself closely to his duties & graduate with honor to himself & family- I hope to have The pleasure of seeing You some time during The summer when I shall be in Connecticut.

Yours most truly
Wm. Turnbull Jr
Baltimore May 20th 1853

To Mr. John Ryan

Manayunk May 31st 1853

Dear Sir

We have the several favors you sent us by the hands of your son Geo. and note their contents. We have procured a lease of the premeses we spoke of on very favorable terms. We have as much power for $600 a Year as we paid $1725 a year for during our first three years in this place. We conclude that we could clear $100 a week on our sett if we were running at the present time.

We are very much pleased to be able to state that friends have come forward to our assistance that will enable us to pay nearly, if not quite, the whole cost of the machinery when finished. It is we presume useless to tell you that we are extremely anxious for the machines, and that we trust you will loose no opportunity to urge on their completion. With reference to the size of the buring machine, we have only to say that we will be satisfied with the cards, being the same as Bruner's in every respect throughout. We would take this opportunity to say that we look on young

Geo. as an exceedingly promising young man, and that we believe he will be entirely successful and prove a credit both to himself and to you.

Yours very Respl^{y}
Reed & Blounts

To Mr. John Ryan

West Point N Y July 3rd 1853

My Dear Mother,

The regulation being to have our coats buttoned all of the while, I find that my present lot of shirts and collars are not very handy besides being much more costly. Therefore I think that if you would make me 6 or 8 shirts of the heaviest cotton you can get (not drill because that would be to rough) they would answer for better.

Make them to open sufficiently far to be easy to take on and off, with 2 buttons in front & one the binding and the other about 1/2 way between the binding and the bottom of the opening, and also one on each risband, the risband being rather short so as not to be visible. If I can get a few of that sort I can save my present shirts for my furlough year. Make them similar to those I wore a few years ago or at least as simple as possible.

I will go out and try to get a sample of the collars, the trouble with those being that they do not touch behind. We wear turn down collars. I exchanged off both of my coats, my pants and my vest for other clothing such as I shall need here. The clothing I got is worth some 30 odd dollars, which I thought was better than keeping them as I should have to lock them up in my trunk and let them stay there and they would in all probability get moth eatened and spoiled. I exchanged with those cadets who were found deficient. This evening the wind blows like every thing and every blast seems almost to carry our tent with it. Hoping you are well.

I remain your Aff Son

George

P.S I suppose that I could get the clothes here but I am afraid that if I am not economical and saving that I shall not have money enough to go on fulough with, provided that time should ever arrive.

To Mrs J Ryan

West Point, N. Y. Aug 28th 1853

Dear Father & Mother

I wrote a letter to you last Thursday informing you that I had been sick but that I was much better. I have been growing better ever since and I am at present able to go to the mess-hall for my meals. I am still rather weak but I am in hopes that I shall be able to return to duty on next Tuesday, the day which we strike tents and leave camp. I was up to Camp this morning and saw quite a number of the furlough class who have returned from their vacation, and the remainder will be here by 2 O'clock this afternoon. Last night must have been a awful night in Camp, as it rained and blowed terribly.

During the week that I was confined to my bed, I would have given anything to have been at home. It made me wish that I had never seen West Point, but as soon as I got out of bed, I began to feel contented again. On the whole I think that I had better be sick now than two or three weeks hence, for, as it is now I have lost nothing but what I can easily learn at any time, while two or three weeks in the beginning of my studies would be enough to alter my standing very much.

I notice by the "Hartford Times" that you had a fire across the road and that Dr. William lost his barn and came very near losing his house. I suppose that you all turned out with your water pails and that you finally had a chance to display your good qualities as fire men as you have turned out a great many times before but never had a chance to operate.

They have "muster" once in two months, at which time our accounts are settled and the ballance of our accounts wether for or against us is carried to a new account. And those who are very much in debt are not allowed to purchase anything until they get out. The muster will ~~probably~~ be on Tuesday or Wednesday.

I hope that when I make any more such bad mistake in spelling I hope you will write to me for in the future I mean to be more correct in spelling. Tell Matthew and John that I will write to them in a few days, tell them also that I would like to hear from them. I think that I may write a little letter to Frank and enclose in the same envelope with this, because my postage stamps are getting rather low. It has been so short a time since I wrote the last letter that I will be obliged to give you rather a short one for this time. I hope that you are all well and that Benny has by this time got well. Give my love to all the family. I shall probably write again to some of the family next Sunday.

Your Sincere and Affectionate Son

George

To John & Johanna Ryan
Norfolk, Conn.

West Point, N.Y. Sept. 12th 1853

Dear Brother,

I received a letter from home, 29th instant, and failed to answer it as generally mean to do on the next Sunday. I was very busy last Sunday and as that is the only time that I can find to write at present, I was obliged to postpone it until today.

I forgot to state in the last letter that we were obliged to go to the Adjutant's office and tell our parent's circumstances, their business and wether they live in the city or country. I told them my father was in moderate circumstances, was a manufacture and lived in the country. Every one I believe said that their father was in moderate circumstances. I saw Gen. Scott a few days ago in the library. He was dressed in citizen's clothes and had a cape on so that I could not see his form so well as if he had his uniform on. He is a montrous tall man and very large. He looks very old and carries his hand in a sling but he still retains the military walk. He was with Col. Lee. Col.

Payne and Judge Wade and appeared like a giant by their side.

I received a letter from H. P. Lawrence the same day I received the one from home and also one from Geo. Bigelow on the next day, both of which I intend to answer in a few days.

I am mighty glad we have left Camp for I had got almost tired of walking post. We are kept here very close in barracks. We are obliged to study and recite 10 hours a day, drill and parade two hours, police and regulate room in the morning half an hour, which with the time we are at the mess-hall does not leave much spare time to walk about. We are at present studying grammar, Algebra and fencing. While we are reciting we are obliged to stand on the floor in the position of a soldier and when we get through and take our seats another gets up an so on. No whispering or playing is allowed in the recitation room. Each section is marched to the Section room and every Cadet not present is reported.

I did not receive a letter from any quarter last week and as I have wrote one letter before this day (to M. C. Ryan) I do not think a very long one ought to be expected from me. Give my best respect to all my friends and my love to all the family. I must look over my lesson for tomorrow and therefore will give you good-bye for another week.

Your Aff. Brother

George

To M. J. Ryan
Norfolk, Conn.

West Point N.Y.

December 31st 1853

Dear Sir

I reply to your letter at the earliest opportuniy as sickness has for several days prevented my attendance upon my duties. I am happy to inform you that you have no reason for anxiety on your sons account. His general conduct is very good

In his studies he succeeds very well. No standing will be assigned to him until after the approaching examination. In his Mathematics he is probably the best in his section which will give him a standing above the middle of the class. In his English studies he is not as well but improves daily and is in no danger whatever of being found deficient. He appears to be ambitious and with the same industry and attention continued he will be sure to attain a very respectable standing in his class and to go through the Academy with credit.

I have no appropriate title or address except that of Professor

Very Respectfully
Your obet Servt
A. E. Church

To John Ryan Esqr.
Norfolk Ct

In a letter dated June 6, 1853, George asked his father to send along a written request to have him excused from the Protestant services held on Sundays in the Cadet Chapel. Superintendent Col. Robert E. Lee accepted John Ryan's request, and George was allowed to attend a nearby Catholic Church, which he did "once a fortnight." At the time, there were fewer than 10 Catholic cadets at "the Point."

West Point N.Y. March 23rd 1854

Dear Sir

I received your letter yesterday and in compliance with your wish I will endeavor to give you all the information you desire.

I think you are mistaken about your son's having 84 demerit for January, he has but 86 demerit for the whole year, that is from the first of July up to the present time, taking all things into consideration I do not think that is a large number, on the contrary it is doing very well indeed. The authorities are much stricter this year than formerly, which makes it more difficult to keep from getting demerit. I find that all your son's demerit are for violations of minor Regulations such that every Cadet is liable to overlook, no sign whatever of bad conduct. I think that when you see the register next June you will find that your son has done much better than the majority of Cadets, many of them have already nearly two hundred. I know from experience that it is a very difficult matter for a Cadet to keep from getting demerit. I myself have a larger number this year than I have had since I have been a member of the Academy

I have had a very good opportunity of judging your son's conduct as he is a member of the Company of which I have command. I have, on all occasions, noticed that he conducted him self as well as any member of the Company. I have never had occasions to speak to him more than once or twice during the year. and then it was for some trifling mistake. I do not recollect of having seen him neglect his duty, he performs his duty as well as any member of the company. It would be impossible for me to tell you all the things that demerit are given for. I dont think any one can understand them or learn them until he has been a member of the Academy, he will then see how difficult it is for one to go through here with out demerit. The very best Cadets in the Corps get demerit as well as the others, the authorities do not consider it as any sign of bad behavior.

I think you may quiet your fears about your son's conduct, if you were only here your self for a few months you would soon see that he conducted himself very properly.

I do not know the reason you did not receive the report for the month of January. I suppose it must have been some mistake made at the War Department

I do not consider that you have taken any liberty whatever in addressing me, on the contrary I shall always be happy to give you any information that I can. I will talk to your son about his demerit and give him all the advice I am able to with great pleasure. I hope you will not be at all frightened about him

Very Respectfully Yours
Charles N. Turnbull

To John Ryan Esq
Norfolk Conn

GR

Providence Rhode Island

June 3d. 1854

Dear Sir,

A little time Since I met the clergyman at "Cold Springs" opposite "West Point" and charged with the congregation at "West Point" and recommended your Son to his special attention.

He informed me that your Son was not attentive to his religious duties, as also that he was not a regular attendant at Mass on Sundays. He informed me moreover, that he spoke with your Son on the matter of his inattention but without effect. This clergyman is out of health. I do not remember his name, and is, at present, either aided or Superseded by another.

You will do well to encourage him, without appearing, for the present, to be informed of his neglect in the matter, to receive the Sacraments, as also to be regular in his attendance at Mass, on Sundays.

When at N.Y you would do well to visit West Point and, with your Son, the clergyman at West Point, and charge the Son in his presence with attention to his religious duties. The Catholics at West Point are poor, and Catholic Cadets few, So that there, love for religion can be the only attraction to its practice. Interested as I am in every thing that concerns you, I felt it proper to give you this injunction, Satisfied that you would do in the case, all that a good father could do.

Most respectfully yours
+ Bernard OReilly, Bp of Hartford

"I noticed that you improved in penmanship and if you continue to make as much progress in proportion you will be a superior writer and that will certainly be a great recommendation as you know that Father has always said that if a person could write well it was much easier to succeed as a clerk."

--George Ryan in a letter to his brother Mathew, Dec. 9, 1855

Litchfield July 25th 1855

Dear Sir

I am glad to hear that you are enjoying a visit from your son. I am expecting to remain at home during the present and next weeks, and will be happy to see you here at any time that may suit your convenience.

Yours very respectfully
O. S. Seymour

To John Ryan Esq.
Norfolk Conn.

West Winsted Nov 19 1855

Dear Sir

Knowing from experience, how acceptable may be a word of sympathy in troubles such as yours, I cannot withhold an expression of my deep regret, in hearing that the money kings have come down upon upon you. Thank God, they cannot sweep away, with the property, the good name of such hard working man as you and your partners. This, I feel assured, remains, a sure investment -- a mine of treasure, worth more than the accumulated millions of the richest money changers in the land.

With sentiments of sincere regard I am, in haste

Truly Yours
John Boyd

To John Ryan Esq

Confidential

New York 11th Mo. 1st, 1855

[To] J & E. E. Ryan & Co

Respected Friends

Your recent favor is received and takes us by surprise. We have relied upon you looking out for our interest however much you might be embarassed and we may truly say that upon this faith in your securing us in any event have we delivered property to you which you know is according to your respected assurances to us and we cannot but think you yet intend to do so by us -- You ought to have consigned goods to us to cover all your indebtedness due us -- You can do this yet can you not?

Your assured friends
Murray & Davis

West Point Nov 22d '55

My Dear Father and Mother

Your affectionate letter of yesterday
came to hand by this day's mail conveying
the sad intelligence of your late misfortune
which pains me much as I was not expecting
anything of the kind although I knew that
business was quite dull still I was hoping that
you would be able to stand your hand.

Such not having been the case I can readily
conceive the grief and sorrow which must
surround the family at home and the thought
of the lonelyness that is about the fireside
causes me the greatest distress and although I
hope all may yet come out right still I can not
help thinking of the distress which present
affairs must cause you. I can not exactly
understand how you have arranged business.
It would seem that you are going to sell out
everything for the benefit of a few principal
creditors and that you are to run the works
for them the profits going to pay you debts to
them.

I then understand that the property is to remains to the new owners weather there be profit or loss and that in case of no profit your time in there business will bring you nothing and judging from what has passed I can not see what prospects there will be for gain and as it will certainly cost you something to live it appears to me that you will be going in worse and worse. I hope that I may have a wrong idea of it and that your next letter will make things more clear which you promised to write in a few days.

In the mean time I hope that things will turn out better than you anticipate. Often when things look so bad at first, they change to the better when investigated and such I hope will prove true in your case. Although your troubles must be very great I hope the worse has passed away with the first outbreak and that you will be able to proceed without much hinderance.

I wish you would write me by Sunday if you have time as I am very anxious to know how you get along as I am afraid that you will have some trouble with those men. Trusting for the better

I Remain Your aff Son

George

Bishop Bernard O'Reilly visited Europe in 1855 and embarked on his return voyage in 1856 aboard the SS Pacific, which sailed from Liverpool for New York. The ship, however, sank off the coast of Wales in the Irish Sea and Bishop O'Reilly was never heard from again.

GR

The Old Academic Building

George Kensel

Francis Beach

West Point Feb 8th 57

My Dear Brother

I received no letter from you on last Sunday but the week before I received two the snow having delayed the previous one. Two weeks to day I wrote you but imagine you had some difficulty in reading it as I was then confined to my bed from an accident which I received while at riding three weeks ago yesterday.

In the new riding hall the course is covered with tan bark to make the ground level and also to give the horses a good footing and it is always necessary in the fall and summer to throw on water to settle the dust. The cold weather has frozen this water so that the hall is very hard and extremely slippery. While cutting heads one day our instructor wishing to give us a good ride formed us in columns of twos a few minutes before dissmissal I coming in the second set of twos and on the outside. As there was but a few minutes left and all were anxious to make the most of it the men put spurs to there horses and we went around there in fine style but at a fearful rate. Finally in turning a corner my horse fell and to the inside thus leaving me at the mercies of both ranks and we were going so fast and being at a turn the men in rear were not able to stop instantly and two set of twos rode over me. The fall itself did not hurt me much but the other horses trampling on me bruised me a considerable. The principal bruises were on the legs and hip but not much considering the circumstances. I went to the Hospital and for the first eight or nine days was confined to my bed and some of the time pretty painful.

To day I left the Hospital and tomorrow I shall commence going to recitation and will probably be able to attend all duty in a few days. The extreme cold weather which we had when I last wrote has all passed away and at present we have quite a thaw. I noticed in a Washington paper a few days since the arrival of Woolcott and Northway and should not be at all surprised if it has something to do with the Cadet appointment. Do you know whether Dr. Welch is mixed up with the corruption case at Washington? I thought I saw something of the kind in the paper. I have not heard from Manayunk since John left as it is not a great deal of pleasure to write and never receive any answers. I thought however you stated in a former letter that Father was going on there. I hope I shall hear from you on Tuesday and about the meeting of the directors of which you spoke in your last. Hoping you are all well

I am Your aff Bro

George

When George was 12, he lost two brothers, Charles John, 1, and Thomas Michael, 3. The deaths, which occurred in 1849, must have had a profound effect on George, who seemed to relish his role as big brother.

West Point June 8th '57

My Dear Brother

Your kind letters of Sunday week and two weeks reached me but for lack of news I have failed to answer them before today. I should have written yesterday but knowing that the examination would finish today I thought I would wait till this evening.

Today the examination of the first class was finished and we will have no more to do although we will not probably be able to leave West Point before next week - at the fartherest however I expect to be home on Sunday week. The standing in all the studies have been published except Infantry Tactics & cavalry Tactics. My standing in Eng. is 33, Ethics 28, Mineralogy & Geology 33- Arty. Tactics 33- I will have plenty of money to pay all my expenses home and such debts as I propose paying at present. I have been introduced to Mr Deming, one of the Board of Visitors and am very much pleased with him. He has been chosen to deliver an address before the class and I am in hopes he will acquit himself well.

Mr Ferris my Plebe reported last week and from what I have seen of him I am very well pleased with him. He appears to be a right smart fellow and should think he would get along quite well but this however is all supposition as it is utterly impossible to say how any man will succeed. Looks are very deceitful as many men come here whom you would think could not stay at the institution for six months yet they graduated well while others who had just the contrary appearance were sent off at the first examination. Hoping to meet you all next week in good health

I Remain Your Aff Bro

George

FIRST CLASS—38 MEMBERS—1857.

Order of general merit.	Names.	State.	Date of admission.	Age at date of admission. Years.	Age at date of admission. Mos.	Order of merit in Engineering.	Order of merit in Ethics.	Order of merit in Mineral. & Geol.	Order of merit in Infantry Tactics.	Order of merit in Artillery.	Order of merit in Cavalry Tactics.	De-merit. For the year.	De-merit. For the 6 mos. ending June 6th, 1857.
*1	John C. Palfrey.	Mass.	July 1, '53,	19	6	2	1	5	2	1	2	6	6
*2	Richard K. Meade, Jr.	Va.	do	16	10	1	2	2	1	2	1	16	12
*3	E. Porter Alexander,	Ga.	do	18	1	4	5	4	3	3	4	16	12
*4	Henry M. Robert,	Ohio.	do	16	2	3	3	3	6	4	8	64	38
*5	George C. Strong,	Mass.	do	20	0	16	7	8	4	12	3	0	0
6	J. L. Kirby Smith,	At Large.	do	16	11	9	4	1	5	5	12	145	85
7	Thomas G. Baylor,	Va.	do	16	2	6	6	7	10	6	7	-87	30
8	Haldimand S. Putnam,	N. H.	do	16	8	5	8	6	7	7	9	46	21
9	William P. Smith,	Va.	do	20	2	7	13	9	9	9	6	30	14
10	George A. Kensel,	Ky.	do	16	11	11	12	10	8	13	14	18	12
11	Thomas J. Berry,	Ga.	do	17	9	8	10	14	19	16	5	111	50
12	Charles H. Morgan,	N. Y.	do	18	8	13	14	17	14	17	21	90	54
13	Oliver H. Fish,	Ky.	do	19	6	17	23	18	13	23	17	27	9
14	Abram C. Wildrick,	N. J.	do	16	11	18	16	16	12	10	15	80	44
15	Charles J. Walker,	Ky.	do	18	3	15	18	19	11	14	10	46	40
16	Francis Beach,	Conn.	do	18	2	10	9	11	18	19	22	184	94
17	William Sinclair,	Ohio,	do	18	4	12	17	25	21	15	13	40	16
18	Augustus G. Robinson,	Me.	do	17	11	20	15	27	20	22	34	146	77
19	Samuel W. Ferguson,	S. C.	July 1, '52,	17	7	23	19	13	16	11	16	183	89
20	Marcus A. Reno,	Ill.	Sept. 1, '51,	16	9	14	11	12	22	8	20	179	81
21	Edward R. Warner,	Pa.	Sept. 1, '53,	17	11	19	22	20	23	20	23	68	32
22	Manning M. Kimmel,	Mo.	July 1, '53,	20	8	26	27	28	30	28	25	107	73
23	George H. Weeks,	Me.	do	19	5	24	24	21	28	30	29	82	49
24	John T. Magruder,	At Large.	do	16	5	21	26	24	29	24	28	125	69
25	George A. Cunningham,	Ala.	do	17	0	30	20	22	31	18	26	144	62
26	Harry C. McNeill,	Texas,	do	18	5	29	25	29	17	32	11	161	80
27	Ira W. Claflin,	Iowa,	do	19	0	35	21	26	15	29	24	103	81
28	Aurelius F. Cone,	Ga.	do	17	0	25	30	38	26	27	30	173	88
29	Paul J. Quattlebaum,	S. C.	do	17	0	22	36	23	34	21	27	93	61
30	Johh S. Marmaduke,	Mo.	do	20	3	32	31	30	35	35	18	151	78
31	George W. Holt,	Ala.	do	18	0	31	33	32	32	34	31	107	21
32	Joseph S. Conrad,	N. Y.	do	19	10	27	37	15	25	25	33	174	81
33	Edward J. Conner,	N. H.	do	18	0	37	29	35	27	31	32	153	88
34	George Ryan,	Conn.	do	17	2	33	28	31	24	33	36	146	73
35	Robert H. Anderson,	Ga.	do	17	7	36	32	33	33	37	19	131	70
36	Charles E. Farrand,	N. Y.	July 1, '53,	18	3	28	38	36	36	26	37	50	28
37	Thomas J. Lee,	Ia.	do	18	3	34	34	34	37	38	38	138	72
38	Lafayette Peck,	Tenn.	do	20	0	38	35	37	38	36	35	53	12

GR

What a horrible fire have happened in our
family - at that time not one of us would
have believed it if we had been told it.
— George Ryan, Dec. 21, 1858

II

Norfolk

Some of the Norfolk folks pretend to regret - your absence and think you had better stay about here - it was a pity for you to leave, &c, &c. You however understand their gammon and know how much it is worth. I trust that we shall all of us soon be off and for my part - I shall shed but few tears in rememberance of Norfolk or its "Codfish Aristocracy."

— from a letter written by George's brother Mathew Ryan to his father on July 18, 1858

Newport Barracks Ky

May 4th 1858

Orders No. 56 (extract)

II. Having turned the Detatchment over to the Commanding Officer at Fort Levenworth the above named Officers will return to this Post except 2nd Lieut. George Ryan, 1st Infy, who will proceed to Jefferson Barracks Mo and report to the Commanding Officer of the 7th Infy in obedience to Special Orders No. 63. dated H^{d} Q^{rs} of the Army New York 26. April 1858.

By Order of
Bvt. Lt. Col. R.C. Buchanan
(signed) H.B. Hendershott
1st Lieut. 2nd Infy
Post Adjutant

Ft Leavenworth K.T.
May 30th 1858

My Dear Parents

Tomorrow we will take up our line of march and although I have a great deal to do I take the advantage of a few moments leisure to write you a few lines- I wrote John last week and I have also written home once before since my arrival but as yet have not heard one word from any of you. I suppose however the reason is that you did not know at first how to direct your letters. We were to day inspected by the Inspector General and the orders are to leave this post at 7 a.m. tomorrow and although I hardly think we will get started so early as that still there is no doubt but that we will get on the road supposing we go but a few miles.

Our column- commanded by Col Morrison & consists of six companies of Infy. a squadron of Cavalry. a company of Drags. a Light Battery. the ordinance and Topographical departments. The chaplain of the Utah Forces also goes in our column. He is Father DeSmith who has been a missionary for so many years in the western country. He has a large wagon to himself and I have no doubt but he will travel with as much ease as the Gen. himself. From what I see. I should not be at all surprised if it was some time before I returned to the States. It seems to be the general opinion that the 7th is destined to stay either in Utah or that section of the country for some time to come- say three or four years.

I send you my pay account for July which you can get cashed in New York any time after the 1st Aug. When on the plains I shall probably have none or at most very poor opportunities to send money so I thought I would take this way. All you have to do is to go to the Paymaster's office No. 6 State St. N.Y. and give him or his clerk the account and they will hand you the money. I wish to have you give ten dollars- $10 of it to Mat. I borrowed that much of him in Phil[d] and have not yet returned it. When I have an opportunity I shall send you letters from the plains. and I would write you more this evening but that I have been so busy that I am quite tired and I also wrote John a few days since giving him all the news. Give my love to all and remember me to my friends. Hoping you are in good health

I am Your aff Son

George

To
John & Johanna Ryan

Norfolk, June 27th, 1858,

Dear Father

By the time this letter reaches you I suppose you will have seen something of Pana and have informed yourself somewhat of the advantages of the State of Illinois to a man of your Proffession. I trust that the result of your experience will be as favorable as we have anticipated. The country, customs of the people &c I suppose are different from those here and I hope that they will be more to your taste. From all accounts which I hear, things are very dull all though the west and so long as grain remains at as low a price as at present I suppose they will be apt to remain so- however I should not think it would affect your business very much. I have heard so much about the west and so many conflicting statements in regard to it that I place no reliance upon any of them: but wait to hear the result of your experience, which I trust will be satisfactory to you.

We have had no letter from George since you left - I think I will write to him to day. No letters from Charley or John this week. Uncle Mathew and myself will go to Litchfield on Tuesday as the court sets in that day and Hogans case is one of the first to be tried. Stevens and Haley are also going down on that day. I saw Stevens yesterday and he seemed to expect a favorable result which I hope will be the case. Haley expects to go to Pana in about a month he told me. It would seem as if Norfolk would be well represented in that quarter.

As regards business I have nothing new to state. We have heard nothing from Lane or Ransen since you left, and I think we have not had a letter from any source. I can see no prospect whatever of anything being done here and every day makes me more convinced that the present concern will never do anything here.

People about here seem supprised to learn that you have gone, most of them I suppose thought that you had no intention of leaving. All who I have heard say anything seem to think that you will do well.

We have had no calls for water yet - the stream seems to hold out pretty well so far but a few more days such as we have had lately will make the water pretty scarce.

There will be no celebration of the 4th of July here this year, which is nothing uncommon for Norfolk. The Winsted people are trying to make an excitement and draw a croud, There is to be a Balloon ascencion &c. I hope you will have a very pleasant day among your new acquantances, on that day.

I suppose that you are somewhat lonesome at times but not very much I hope. You must keep your spirits up as much as possible and you must not worry at all about the family as we are getting along very well, full as well as I expected.

From all I can learn money is very scarce in all parts of the west and commands very high rates.

We shall be expecting a letter from you soon and shall hope to be more fully informed of the advantages and disadvantages of the western country as they appear to you. I suppose it is a very hard time to do any kind of business and after thinking over all manner of business, occupation &c it seems to be that whether the times are good or bad that a lawyer stands as good a chance of making a living as most any other person.

I hope you will like the country, the people climate &c and that your expectation (which were never of the grandest kind) will be fully realized. Above all things I hope you will find a more liberal sentiment than prevails in these parts. Is the Democratic party the ruling one in Christian County? I hope you will have the good luck to locate in a Democratic section of the country.

You must give yourself no uneasiness about any of us at home as we are all quite well. Uncle Mathew and family are all well. He predicts that your career in the west will be success full, more so than I guess you even thought.

Uncle Charleys family are also all well. We are all well and join in sending you our love and hope that this will find you in good health and enjoying yourself and that we shall hear from you soon. I remain

Your affectionate son Mat

Litchfield July 13. 1858

Dear Sir

Yours of the 7th inst. was recd. last evening. I had not before heard of the change in your location & business. I wish you success in your new location. You are at liberty to refer to me for character & it will give me much pleasure to speak in your favour to all whom it may concern.

Yrs very respectfully
Origen S. Seymour

Norfolk September 17th, 1858

Dear Brother John

Your Mathew. has informed me he received a letter. from you. saying an Irish Lawyer in good. practice. from. Springfield Illinois had made you an offer. to take you into. partnership. and give you half. the profits of the business. that he was a good Lawyer but an intemporate man. you ought to know your own business best. but it would seem to me. it is just the offer. you want. with your large. family. and small capital in money. the chance of. a Lawyer beginning business. and his earnings. is at best very precarious while on the other hand. you have nothing to loose and have a good chance. to. improve yourself in the Law, there may be some small objections. but I think. very. favorable of going into partnership with him. however I now but little about your situation compared to yourself. and you must be your own judge, it is my anxiety for. your welfare and the settling of your. family that induces me to intrude into your business. but I hope my good intentions. will excuse my freedom as I wrote you a few days ago I have nothing. new, after the Quarterly meeting comes off I will write. you. hoping you are in good health and that the Almighty God. will bless your labour

I Remain
your Broth Mathew Ryan

In its original meaning, the term Puritan meant "those who strove for a worship purified from all taint of popery." Emerging from the shredded cloak of the Protestant Reformation, Puritans rejected Rome and the seven sacraments of the Catholic Church, opting instead for a congregational, almost civic approach to God and the Gospel.

Norfolk, Sept. 19th, 1858

Dear Father

We are getting in shape to leave, as fast as we possibly can, and have, thus far succeeded better than we expected. We have sold nearly all our goods, and considering the times, at what I would call fair prices. There are as, a matter of course, many things to be done in the way of getting so many ready such as getting clothing for the childen, which we must have in readiness before we leave, and these we are attending to as fast as possible, and hope to have all things in shape, so that we may leave here by the 4th of October, or very soon after.

We shall give you notice, in due time, as to the day of our departure so that you may know when to expect us. If you should make up your mind to go to Springfield it may break up your plans so much that you would not want us to come so soon, and if such should be the case we can wait untill you are ready for us to come, which I hope would be in a short time as we are all anxious to get away.

It appears to us that it would be as easy and may be more so to find a house in Springfield than in Decatur, and I should think it would be a very pleasant place to live in. We are satisfied to go to any place that you think we can do best in, and I trust that which ever place you choose it will prove to be such a one. We are not in shape to be very particular at present and for this reason it will not be very hard to suit us. You must not make too much trouble for yourself in getting a house - for the present we can get along without a very large one, and when we are better able it will be time enough to get such a one. We will have to get along with as little expense as possible untill we get to have more of an income and when we do we can live a little "faster". It will take considerable to furnish any house that we may get: but if you get a large one it will take more. You must get such a one as you think will answer, and we should be satisfied.

I am very glad that we are going to make the trip and see if we cannot make a living in some other way than by hanging around this place untill we would be completely run out and good for nothing. I always thought and think so still that we can make a good and respectable living and some thing more and I believe that we can, at least we are going to try, and I hope we shall make it out.

As the time for leaving Norfolk draws near I cant write to you half what I would like. I have so much to say that I dont know where to commence; but I have one consolation and that is that we shall soon be together once more and that we will then have an opportunity to talk and say as much as we please, and to arrange our plans for the future to the best advantage that we can.

The neighbors all appear to regret our leaving and seem willing to do any thing for us and have been very kind. None of us however feel very low spirited at the prospect of leaving but on the contrarilly are ready to "try our luck" in the West and are in a hurry to get a way.

Your aff son
Matt Ryan

Norfolk Conn Feby 6th 1859

Dear Mat.

Yours of the 30th is duly received we were all very sorry to hear of the death of Mr Reed. his death must be a great loss to his family. There is not much passing here that would be of interest to you things here in Norfolk follow on in about the same old routine one day some little excitement the next something else.

The principal subject of gossip with the village people at this time is the building of a Catholic church. they seem determined to go ahead with the building. have a building committee appointed and have commenced getting out the timber. Timothy is our leading man in this as well as in politics and every thing else you would laughf to see the importance of Tim he thinks that he is perfection in all his speeches the closing remarks are that he is very sorry that his gifted namesake in Illinois is not present to address them. he made a great speech at the Democratic caucus the other evening. he closed by saying that the gentleman now in Illinois but who was formerly a leader among them might yet hold a seat in the Halls of Congress. (great cheers)

busines about Norfolk does not seam to improve much. the Norfolk Woolen Co are running full now but the more they run the worse off they are. they pay their help very small wages and are every day running down the quality of the goods they have given up using wool and are now back on the old track rugs & Cotton. Some of them are begining to find fault with Isac management it would not surprise me a great deal if he was turned out of office before long.

There is a stockholders meeting called on the 16th for the purpose of electing officers for the coming year. Lawrence is beginning to feel pretty cross because his interest is not paid he says that if the payment is delayed much longer that he will go to work and close of the Mortgage that he has said all that he ever will say about the matter and that to close the Mortgage will be the next step. Isac is thought but very little about by the Norfolk people generaly he has a great opinion of himself and has to take some hard raps on that account. John Nettleton gave him a tongue thrashing a few evenings ago at a Lyceum which they hold once a week in our prayer room I never John so much excited and he never came out more personaly on any than he did on Isac.

I suppose that you wud like to hear how business flourishes with ourselves. I cannot say that our business increases we have done very little for the past two weeks we must either do more or give up however this is a dull time with all the storekeepers in town I think that we get our share of trade

Nat. Stevens has been a hard customer for us he must be owing us some three hundred dollars one hundred of which is borrowed money and it is imposible to get a dollar from him. we had to pay the Woolen Co. three hundred dollars the first of this month which with Nats bill makes money pretty close with us. I have come to the conclusion that we must get credit. some where or else or never do business in Norfolk. you must not think by this that I am in the least discouraged I am determined to stick it out come what may.

I receive a letter from John during the week he says that there is a great excitement about the gold where he is & that he thinks that he will start for Pike Peak. come spring. remembe me to all the family

yours affect.
Chas M Ryan

George was born in 1836, the same year that the first Catholic Mass was celebrated in Norfolk. It was presided over by the Rev. James Fitton of Hartford, and held in the home of Mathew Ryan. Twelve people attended that first Mass, including John and Johanna Ryan. In subsequent years, missionary priests celebrated Mass in the wool-sorting room of the Ryan mill.

Norfolk May 8th 1859

Dear Mat

Your very interesting letter of the first inst came duly to hand We are all very happy to know that you are all in good health. and hope you may always report the same.

I think I mentioned in a former letter that we were about building a Church the work has now commenced the foundation is nearly half finished and I think there is evry prospect of closing in the building this season if nothing more. I take considerable interest in this matter and mean to use all the exertion in my power to keep the thing along, not that I have be come religious or hypocritical (but to tell the truth) more for the purpose of anoying these old Blue about Norfolk they have made their brags all winter that the Paddies & Catholics were all rioted out and that there would never be a Catholic Church in Norfolk. had the blues only kept a little quiet I persume there would have been no church put up this summer. while their talk about the subject has so exerted the Irish that they are determined to have up a Church if for no other purpose they say to spite the blues.

report is going the rounds, now that Peck told Isaac unless business was better managed that they should have to find some other person to take charge of the concern. Since they commenced finishing the goods Murphy has had hard times some ten thousand yards of cloth has been sent back to be finished over. Ranson told father that unless they did better than they were now doing the sooner they stopped the better. I am pretty sure that Isaac will not last many months longer. They have made a settlement with Frank Mills by paying him one fourth of the debt in money and the remainder in satinet at twenty five cents. They also gave Tom Barry a couple of lbs. of flour and twenty five Dolls. on his note. Myself and Isaac have not been on first-rate terms since Uncle Charly left at which time I gave Esq Isaac a piece of my mind and that plain. Since then he has been trying to induce some of the help to stop trading with me and tells them that they can do their trading twenty five per cent cheaper with

Shepard than they can wth me. he has had no converts.

If I only had the means I am satisfied that there could be a good living made in Norfolk by Store Keeping but still have no wish to remain here. Remember me to John if he is with you yet tell him to write me a line. We are all well and hope to find you in good health remember me to all the family. Your afct

Chas M Ryan

The Rev. James Fitton wore a black suit and hat (as Protestant ministers of the day did) when he traveled by stagecoach to Norfolk. But if his true identity as a Catholic priest became known to his fellow travelers, trouble often ensued. Puritan hatred of Catholics was alive and well in the 1830s, and Fitton frequently had to dodge insults and even stones during his mission trips.

GR

Camp near Box Elder U.T. Aug 14th 1859

My Dear Brother

I wrote you a short letter from Salt Lake City saying that I was going out for forty eight days- Our trip passed without anything unusual occuring till the 12th when about five miles south of Box Elder when we had been some two hours in camp- Lt Gay- the officer in command received reliable information that there was a large body of Indians in a valley to the east of Box Elder. He immediately determined to follow them and ordered each man to take rations for four days- in order that he might follow them to the mountains if necessary- and quietly broke up camp at 10 o.c. p.m and moved to this town and leaving his wagons under charge of a Guard and every thing being ready and started at 2 o.c. am the 13th and made a very rapid march over a most humble mountain road and came on a large Indian village at 4 o.c. a.m. and attacked them at once.

The fight lasted from long before day till sun was well up- say from one & half to two hours. We lost no men but had four severely and two slightly wounded and nine horses shot but none of them have yet died. I am think we succeeded very well as the Indians had a most admirable place- their camp being situated at the foot of a bluff so steep that it was impossible to urge the horses but a short distance up it- and on one side a ravine. Our guide a Mormon scoundrel said that there was level ground for three hundred yards beyond their camp and the charge was made at once and the falseness of the guides statement was not apparent- owing to the darkness- till some men nearly precipitated down the ravine and the rest were trying to climb the hill- The Indians at first returned the fire very brisk but it gradually ceased. The guide deserted us in the commencement and we saw no more of him till our return to this place between 9 & 10 a.m. The number of Indians killed is supposed to be about twenty owing to the ground and they always make it a rule to carry their dead from the field. We will probably move forward to day or tomorrow and I imagine that hereafter the emigrants will not be troubled.

The emigration this year is very large almost every day since leaving the city we have passed two or more trains and it is said that at Bear River and where the roads from Ft. Bridger and South Pass intersect this road there are also a very large number.

Box Elder I think the last town we shall see till our return. If however anything unusual occurs express will be sent to Camp Floyd by which I can send letters - one was sent last night - giving an account of the fight - but I was so tired that I was unable to write. This goes by the mail but as the Post master is a Mormon I doubt if it reaches you without being opened. Hoping you are all well and prospering

I am Your Aff Bro
George

To M.J. Ryan
Decatur Ill

Decatur, Ill, June 7th, 1861

Dear Brother,

I have just received your letter of the 3d instant and take particular notice of its contents. As only few minutes is left me to write you, from the time your letter came to hand, until the mail for the East closes, I will have to be very brief in what I write you.

In the first place I will state, that I was strongly of the opinion from the first moment that I heard of the fire, that it was produced either directly or indirectly by Scoville; and further that he made, or rather would attempt to make use of Barny Dunn to do the business. But as to the part I ought to act in the case, I think I have some idea of what it should be. I am a thousand miles from Norfolk, and it is known that I have had a great deal of difficulty with Scoville before I left there and for me to undertake from anything that would be written me to bring Scoville up on such a charge as that of firing or causing to have fired the factory would look too like spite against him. But dear

Brother you are on the spot and can compell any, or every one that you know, that you may suppose know, anything about the business to testify against Scoville in a court of Justice.

Now, Dear Brother situated as you are, one bold and determined stroke on your part may not only save yourself, but would enable the whole of the firm J.& E.E. Ryan & Co. to finally settle up their business with Scoville and the other mortgagees. I am willing to bear my part of the cost of a lawsuit to bring Scoville and the others to terms; but if they are willing to settle without one so much the better.

I would advise you to go and see Richardson yourself, and perhaps it might be well for you to take your son Charles with you and state the facts to that gentleman that you stated in the letter that you have just written to me. In doing that there is no harm, and then tell him you want a final settlement with all parties and be done with the whole affair; and tell him on the terms you will settle; and if not you will take steps to have the insurance money withheld until the matter will be investigated. I would also advise you to go and see Mr. Hitchcock about the business.

Your brother
John Ryan

P.S. I will write you again in a day or two.

Construction on the Immaculate Conception Church (p. 50) was completed in 1859. The church sat "just a pistol shot" from the old Ryan mill, and the suspicious fires in the area must have sent embers swirling all around it.

GEORGE RYAN

NOTES

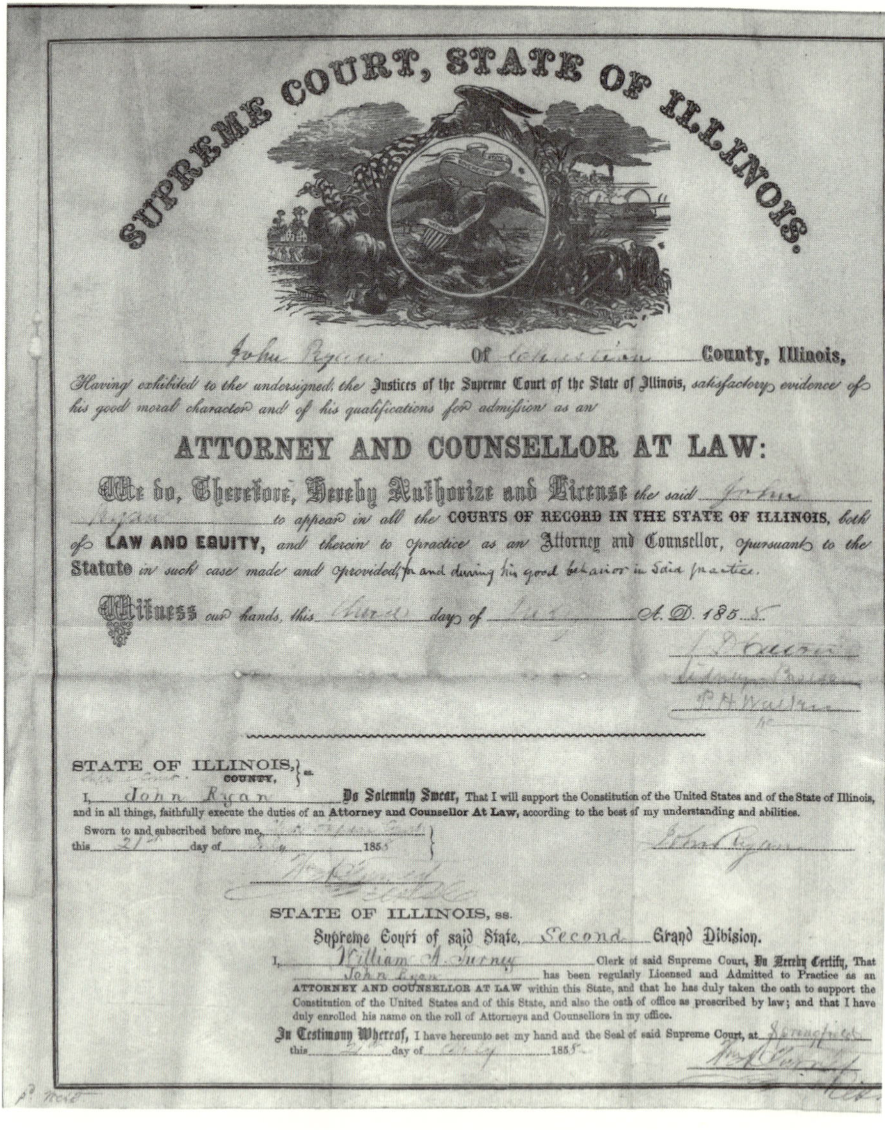

SUPREME COURT, STATE OF ILLINOIS.

John Ryan Of [illegible] County, Illinois,

Having exhibited to the undersigned, the Justices of the Supreme Court of the State of Illinois, satisfactory evidence of his good moral character and of his qualifications for admission as an

ATTORNEY AND COUNSELLOR AT LAW:

We do, Therefore, Hereby Authorize and License the said John Ryan to appear in all the COURTS OF RECORD IN THE STATE OF ILLINOIS, both of LAW AND EQUITY, and therein to practice as an Attorney and Counsellor, pursuant to the Statute in such case made and provided, for and during his good behavior in said practice.

Witness our hands, this third day of [illegible] A. D. 1858.

[illegible]
Sidney Breese
P H Walker

STATE OF ILLINOIS, } ss.
[illegible] COUNTY,

I, John Ryan Do Solemnly Swear, That I will support the Constitution of the United States and of the State of Illinois, and in all things, faithfully execute the duties of an Attorney and Counsellor At Law, according to the best of my understanding and abilities.

Sworn to and subscribed before me, [illegible]
this 21st day of [illegible] 1858

John Ryan

[illegible]

STATE OF ILLINOIS, ss.

Supreme Court of said State, Second Grand Division.

I, William A. Turney Clerk of said Supreme Court, Do Hereby Certify, That John Ryan has been regularly Licensed and Admitted to Practice as an ATTORNEY AND COUNSELLOR AT LAW within this State, and that he has duly taken the oath to support the Constitution of the United States and of this State, and also the oath of office as prescribed by law; and that I have duly enrolled his name on the roll of Attorneys and Counsellors in my office.

In Testimony Whereof, I have hereunto set my hand and the Seal of said Supreme Court, at Springfield this [illegible] day of July 1858.

Wm A Turney

John Ryan was admitted to the bar on July 3, 1858

John Ryan

Johanna Ryan

No. 48

THE PRESIDENT OF THE UNITED STATES

having been pleased to appoint George Ryan

a Cadet in the service of the United States, he is to be received as such, and entitled to all the consideration attached to said appointment. This appointment to take effect from the 30th day of June 185

GIVEN at the War Office of the United States, this 24th day of January in the year of our Lord, one thousand eight hundred and fifty-four and in the Seventy eighth year of the Independence of said States.

Jeffn Davis

SECRETARY OF WAR.

George was admitted to the U. S. Military Academy in 1853

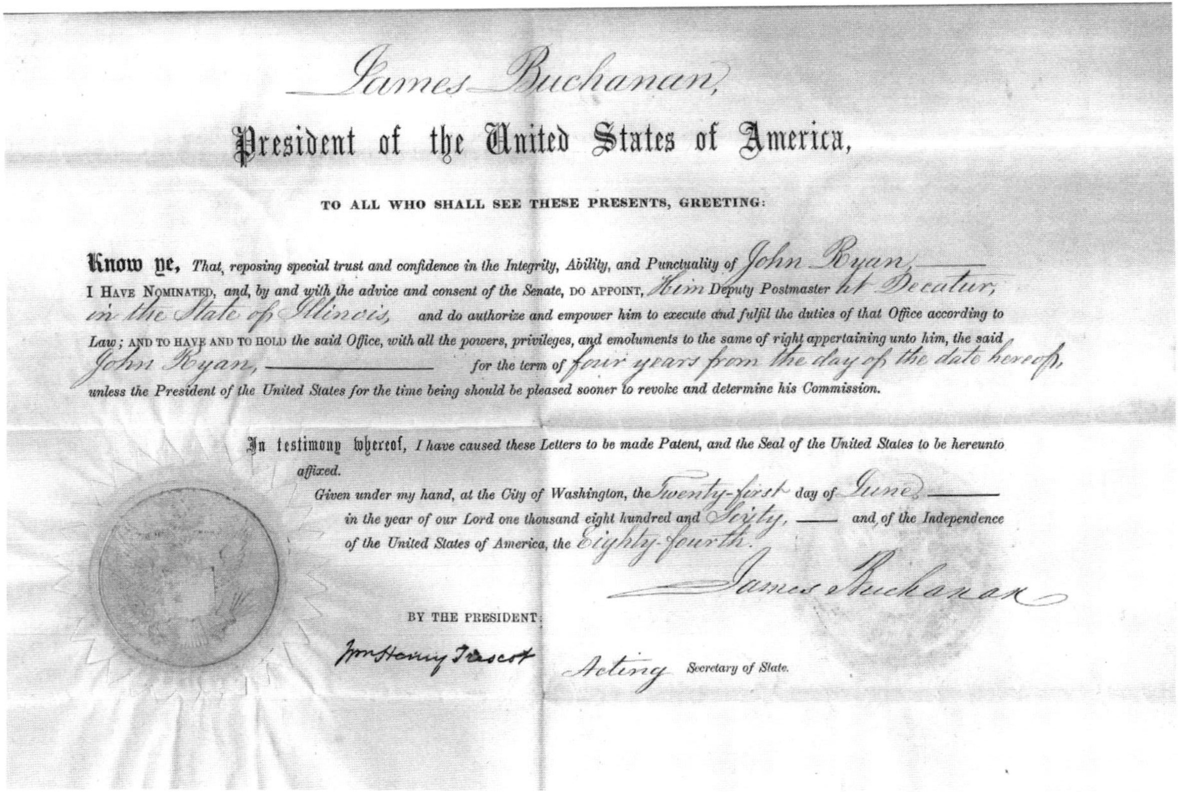

James Buchanan,

President of the United States of America,

TO ALL WHO SHALL SEE THESE PRESENTS, GREETING:

Know ye, That, reposing special trust and confidence in the Integrity, Ability, and Punctuality of John Ryan, I HAVE NOMINATED, and, by and with the advice and consent of the Senate, DO APPOINT, Him Deputy Postmaster at Decatur, in the State of Illinois, and do authorize and empower him to execute and fulfil the duties of that Office according to Law; AND TO HAVE AND TO HOLD the said Office, with all the powers, privileges, and emoluments to the same of right appertaining unto him, the said John Ryan, for the term of four years from the day of the date hereof, unless the President of the United States for the time being should be pleased sooner to revoke and determine his Commission.

In testimony whereof, I have caused these Letters to be made Patent, and the Seal of the United States to be hereunto affixed.

Given under my hand, at the City of Washington, the Twenty-first day of June, in the year of our Lord one thousand eight hundred and Sixty, and of the Independence of the United States of America, the Eighty-fourth.

James Buchanan

BY THE PRESIDENT:

Wm Henry Trescot Acting Secretary of State.

John Ryan is appointed Deputy Postmaster of Decatur in 1860

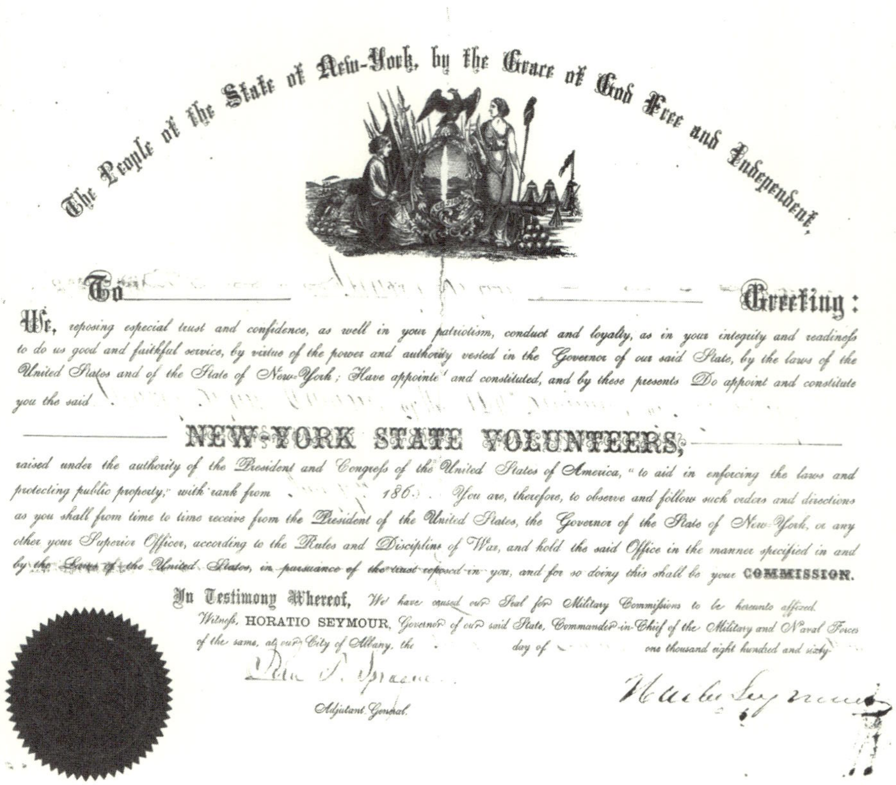

The People of the State of New-York, by the Grace of God Free and Independent,

To ______ Greeting:

We, reposing especial trust and confidence, as well in your patriotism, conduct and loyalty, as in your integrity and readiness to do us good and faithful service, by virtue of the power and authority vested in the Governor of our said State, by the laws of the United States and of the State of New-York; Have appointed and constituted, and by these presents Do appoint and constitute you the said

______ NEW-YORK STATE VOLUNTEERS,

raised under the authority of the President and Congress of the United States of America, "to aid in enforcing the laws and protecting public property," with rank from 186 . You are, therefore, to observe and follow such orders and directions as you shall from time to time receive from the President of the United States, the Governor of the State of New-York, or any other your Superior Officer, according to the Rules and Discipline of War, and hold the said Office in the manner specified in and by the Laws of the United States, in pursuance of the trust reposed in you, and for so doing this shall be your COMMISSION.

In Testimony Whereof, We have caused our Seal for Military Commissions to be hereunto affixed. Witness, HORATIO SEYMOUR, Governor of our said State, Commander-in-Chief of the Military and Naval Forces of the same, at our City of Albany, the day of one thousand eight hundred and sixty

Adjutant General.

George is commissioned Colonel of the 140th New York

Frank Ryan

George and his brothers

John Ryan

Mathew Ryan

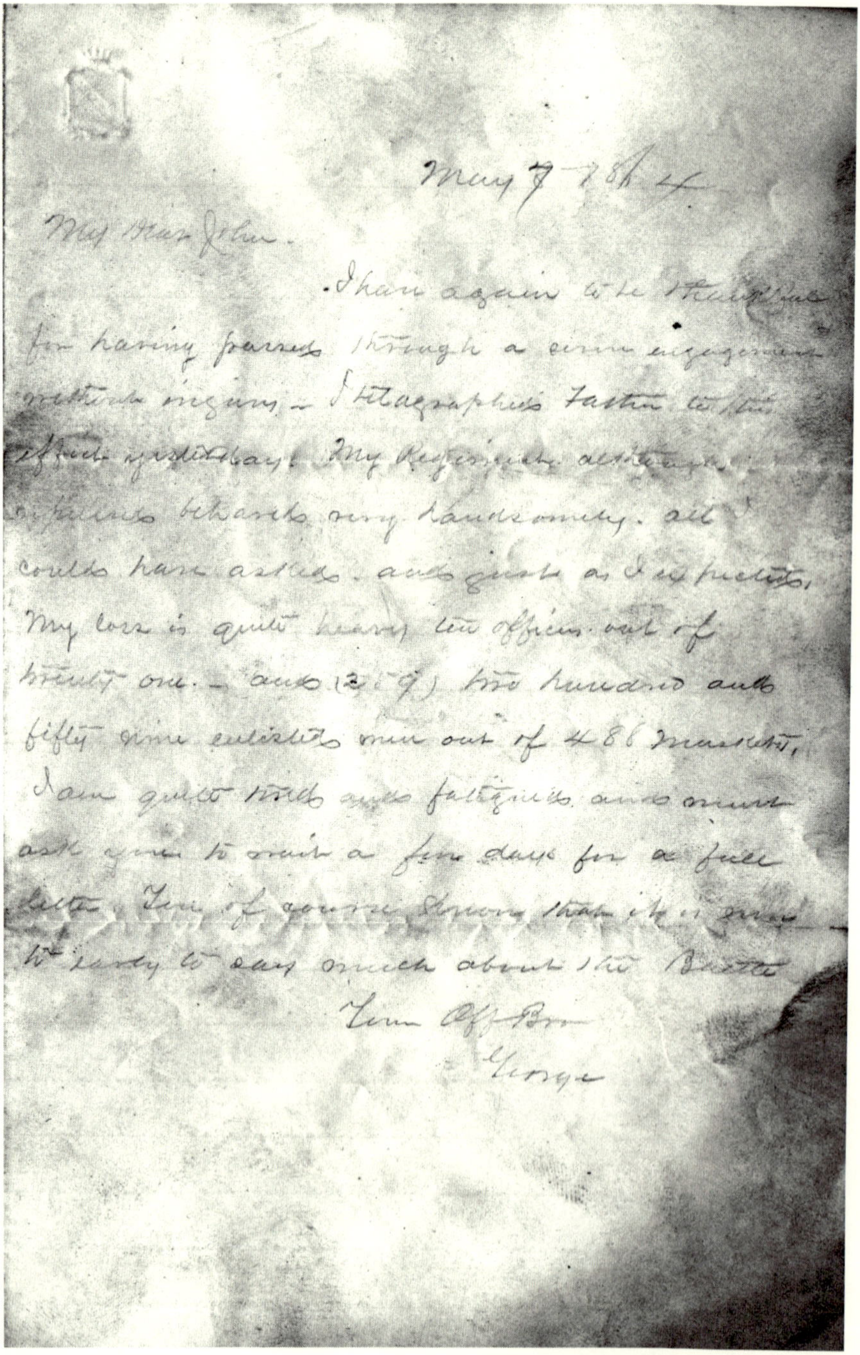

May 7 1864

My dear John.

I have again to be thankful for having passed through a severe engagement without injury — I telegraphed Father to this effect yesterday. My Regiment although engaged behaved very handsomely. all I could have asked. and just as I expected. My loss is quite heavy ten officers out of twenty one — and (259) two hundred and fifty nine enlisted men out of 486 muskets. I am quite well and fatigued and must ask you to wait a few days for a full letter. You of course know that it is much too early to say much about the Battle

Your Aff Bro

George

George wrote this letter to his brother John on May 7, 1864

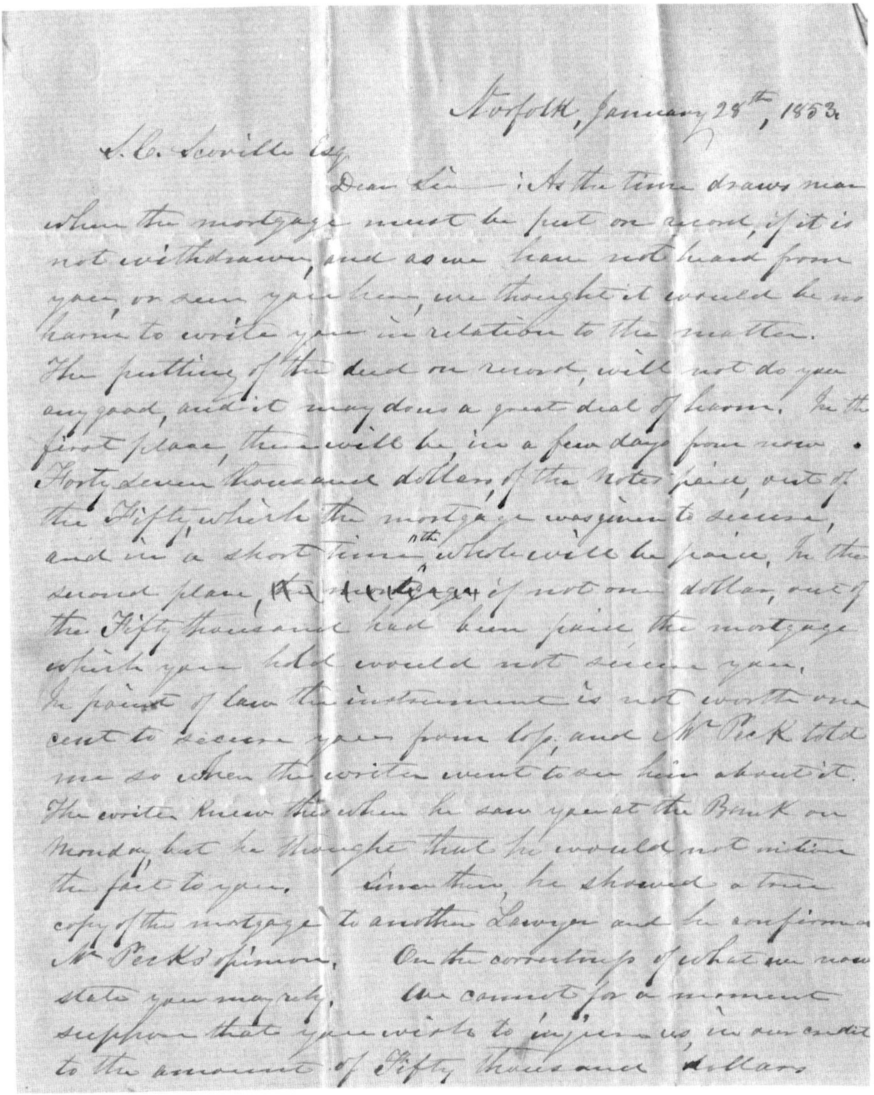

Norfolk, January 28th, 1853.

S. C. Scoville Esq.

Dear Sir: As the time draws near when the mortgage must be put on record, if it is not withdrawn, and as we have not heard from you, or seen you here, we thought it would be no harm to write you in relation to the matter. The putting of the deed on record, will not do you any good, and it may do us a great deal of harm. In the first place, there will be, in a few days from now Forty seven thousand dollars, of the notes paid, out of the Fifty which the mortgage was given to secure, and in a short time the whole will be paid. In the second place, ~~the mortgage~~ if not one dollar, out of the Fifty thousand had been paid the mortgage which you hold would not secure you. In point of law the instrument is not worth one cent to secure you from loss, and Mr Peck told me so when the writer went to see him about it. The writer knew this when he saw you at the Bank on Monday but he thought that he would not mention the fact to you. Since then, he showed a true copy of the mortgage to another Lawyer and he confirmed Mr Peck's opinion. On the correctness of what we now state you may rely. We cannot for a moment suppose that you wish to injure us, in our credit to the amount of Fifty thousand dollars

John Ryan wrote this letter to "the villain" Scoville company on Jan. 28, 1853

This painting of E. Grove and Jerusha Lawrence was painted in their Norfolk home (shown) in 1833

III

Spottsylvania

Mathew J. Ryan found his bride among Norfolk's Protestant upper class. He married Nancy Lawrence, a daughter of E. Grove and Jerusha Lawrence in 1867. The young couple settled in St. Louis, where they raised nine children.

The Ryan Papers, all of the letters and documents in this book, were kept in an old army footlocker in the attic of their St. Louis home until 1970.

For his part, George has spent the last four years fighting Indians in Utah and New Mexico. Captured by Texas insurgents in 1861, he is later exchanged for a prisoner of equal rank, and once again takes up his old war footing.

Decatur, August 30th, 1861

Dear Brother Mathew,

Your very welcome letter of the 19th instant was duly received by my family, during my absence from here. I was on a visit to Mat, who is now in St. Louis. He is what is called a Sutler, that is store keeper to a regiment. If he should not be unlucky, the place will be a good one. This person is not an enlisted soldier, and can leave anytime that he pleases to do so. On my return I hasten to reply to your letter.

You are mistaken in regard to our letters. I wrote the last one, before this now received from you, and yours was realy an answer to mine But this matters little.

I am realy sorry to hear of the premature death of Mr Lawrence's daughter. The loss of a child, at about her age, comes, I think, heavier on the parents than any other. But such things happen and they must be born with.

I have received but one letter from George for a long time and that was two weeks ago.

He then told me not to write to him, until [he] would write to us again. In his letter he states that he was on his way to Fort Filmore, which place it was reported, was about to be attacked. The late advises from that part of the country are that it has been attacked, and by the treachery of the officer in command of it a Major Lynde, a Vermonter by bearth, the whole army there and the post was surrendered. I feel mighty uneasy about George. I am almost sure that he is taken prisoner, and that out in that country would come pretty hard. Let us hope for the best.

As to Scoville I cannot say much to you. You are on the spot and must act for yourself. It does appear to me that with the hold you have on him, through Barny Dunn's testimony, you can make your own terms. Dunn's testimony would not only subject him to a criminal prosecution, but if the Insurance Co have paid the money over, he would be subject to an action from them for the recovery of the same. As for me, at the distance I am from Norfolk I cannot help you, except by way of advice; but I feel very confident that if your lawyer will do your duty you can have your own terms.

You must make up your mind to live in Norfolk, for at your age, and from your habits of life and natural desposition, you must not think of changing your residence. You have a strong hold on Scoville, and never let it go. You may use my name in anyway you may see fit, to frighten him. Even to that, that I want a criminal prosecution instuted against him. I repeat again that you have a death hold on the villain and use it. If you dont you will be sorry.

From all I can learn from the papers the war spirit does not run as high in Conn as it does in the West. Here there is nothing talked of but the war. Illinois has now as large an army in the field as many of the third rate powers of Urope. It must amount to not less than 60000. The New England freesoilers did a good deal to bring on this war and, if I am not very much mistaken, they will get enough of it, before it will be over. Appearances lead us to think that there will be a long war. I would not be surprised if England & France would be mixed up with it.

I am still the Post Master of Decatur but how long I will I cannot say. It would not surprise me if I would be removed any day.

The family are all well and join me in sending you all our love

Your Ever,
Affectionate Brother
John Ryan

Decatur, Aust 28th, 1861

Dear Son,

I have just received your welcome letter of yesterday, and though I do not feel like writing much will write you a few words.

I have just received a very friendly letter from Mr King's son, stating that his father was absent from the city, but that he called to the war department and after much such there found the recommendation in your favor, cast aside among others of the same kind; but on his father's return to the city he would call personally on the Secretary of War. He also wishes me to send him the number of your regiment as that is indispensable in order to have your warrant made out. His charge is only Five Dollars. I thought it would much more. I only received the letters this afternoon and when I take a copy of it will send it to you. In the mean time if your regiment has got a number write me immediately.

I take particular notice of what you say about Mr. Mepham offering you credit. By all means accept his offer. If any bad fortune should befall you it would be much better to be indebted to a merchant, than to Milliken & Odor or Mr Lawrence. I suppose that Mr Lawrence's daughter was in the very jaws of death when he got my letter, and if I had known such to be the case I could never think of asking him for the money. and though we now have it, I would prefer having it on deposit if you can buy goods without it. And further the more extensive your dealings are with Mr Mepham the more chance there will be for you hereafter, if you pay him to get a great deal out of him hereafter. I feel as if I cannot press this matter to much on your mind.

I have not hired any one for you. I will leave that to yourself altogether; but if you want to hire a man for you I will do so.

Jennings for the first time since you left worked on the barn today. I will write you again tomorrow. Write me often if only a few words each time. We are all well

Your Ever Affectionate Father
John Ryan

Decatur, Nov 21st, 1861.

Dear Son,

We have received your favor of the 19th instant with the returned Draft for $50.00. I am glad that you were not in want of the money, for I feared that you were.

A telegraphic despatch was received in this town yesterday stating that Mat's Regt is now on its way to St Louis, and when the dispatch was sent it was 10 miles at this side of Rolla and would reach St Louis about Monday next, the 25th instant. If this information should prove correct you will most likely see him and Frank, and have

an opportunity to talk with them.
I wrote to Mat two days ago, in-
forming him that you were now
at Jefferson Barracks. I direc-
ted the letter to the care of a
mercantile house in St Louis
that he does business with,
as they probably will be in-
formed as to his whereabouts.
The name of this mercantile
firm is M. S. Mepham
& Brother and their store
is No. 81. Second Street.
The first time you will be
in town after you receive
this I wish you would call
there and if Mat should be
in the city you will either
find him at this place, or the
Messrs Mepham & Brother will
direct you where to find him.
I am inclined to think
that Lieut Grey is taken
prisoner. I have not heard
from him for some time,
and some time after he
wrote his last letter to me
there was an account in the
papers of a number of prisoners
being carried to Fort Pinkney
in Charleston harbor and
among them was a Grey an
Army officer. Since this
account appeared I have
received no letter from him.
We are all well, and for
which we feel thankful to
God, and hope this will find
you so too,
Your Ever
Affectionate Father
John Ryan

George was on garrison duty in Missouri when he received this letter from his father

Decatur, Nov 21st, 1861

Dear Son,

We have received your favor of the 19th instant with the returned Draft for $50.00. I am glad that you were not in want of the money, for I feared that you were.

A telegraphed dispatch was received in this town yesterday stating that Mat's Regt, is now on its way to St. Louis, and when the dispatch was sent it was 10 miles at this side of Rolla and would reach St. Louis about Monday next, the 25th instant. If this information should prove correct you will most likely see him and Frank, and have an opportunity to talk with them. I wrote to Mat two days ago, informing him that you were now at Jefferson Barracks. I directed the letter to the care of a mercantile house in St. Louis that he does business with, as they probably will be informed as to his where abouts. The name of this mercantile firm is M. S. Mepham & Brother, and their store is No. 81 Second Street. The first time you will be in town, after you receive this, I wish you would call there, and if Mat should be in the city you will either find him at this place, or the Messrs Mepham & Brother will direct you where to find him.

I am inclined to think that Lieut Grey is taken prisoner. I have not heard from him for sometime, and some time after he wrote his last letter to me there was an account in the papers of a number of prisoners being carried to Fort Pinckney in Charleston harbor, and among them was a Grey, an Army officer. Since this account appeared, I have received no letter from him.

We are all well, and for which we feel thankful to God, and hope this will find you so too.

Your Ever
Affectionate Father
John Ryan

(envelope addressed to)
Lieut George Ryan
7th Infantry, U. S. Army
Jefferson Barracks, Mo.

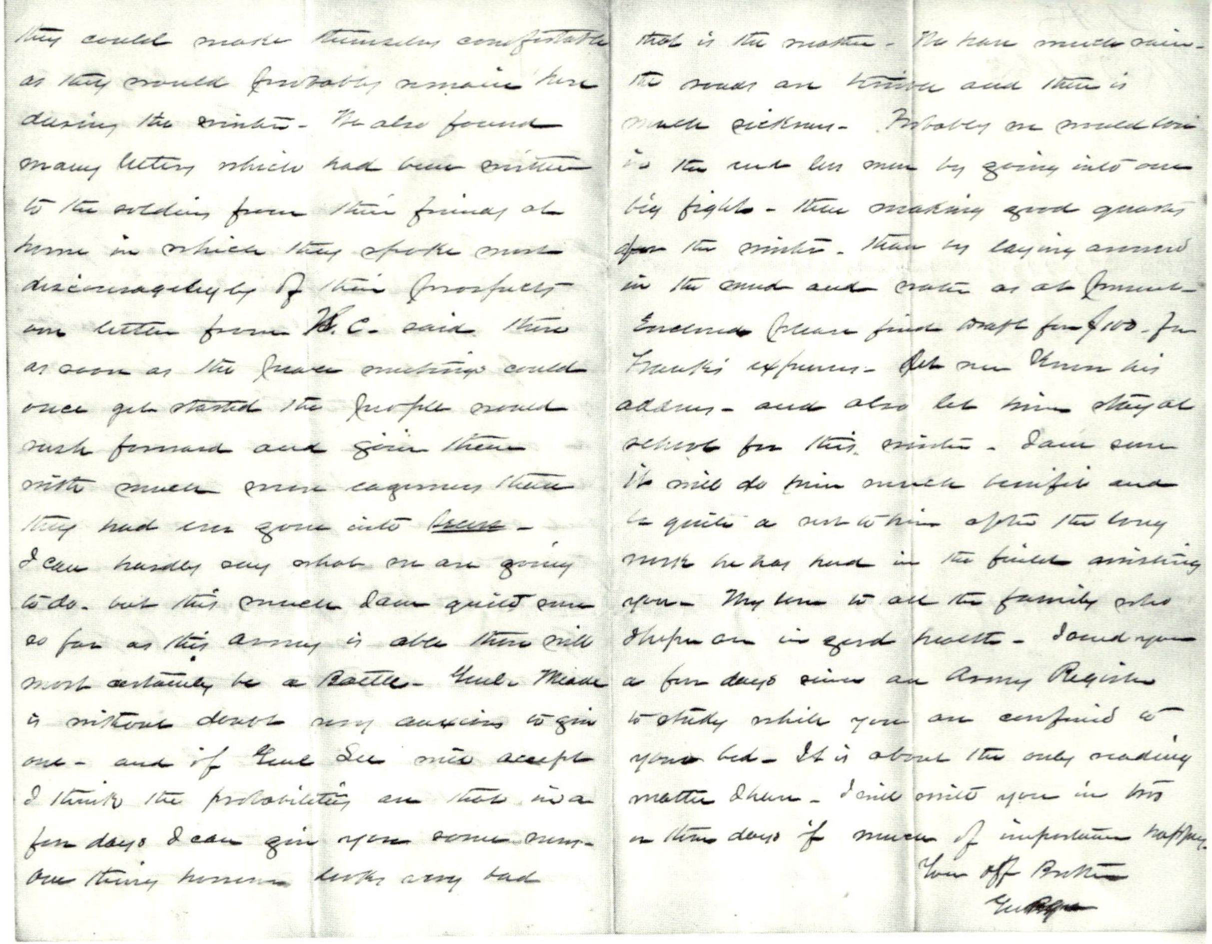

they could make themselves comfortable
as they would probably remain here
during the winter. We also found
many letters which had been written
to the soldiers from their friends at
home in which they spoke most
discouragingly of their prospects.
one letter from N.C. said there
as soon as the peace meetings could
once get started the people would
rush forward and give them
with much more eagerness than
they had ever gone into Secess -
I can hardly say what we are going
to do. but this much I am quite sure
so far as this Army is able there will
most certainly be a Battle. Genl. Meade
is without doubt very anxious to give
one - and if Genl Lee will accept
I think the probabilities are that in a
few days I can give you some news.
One thing however looks very bad
that is the weather. We have much rain.
the roads are terrible and there is
much sickness. Probably we would lose
in the end less men by going into one
big fight - then making good quarters
for the winter. than by laying around
in the mud and water as at present.
Enclosed please find draft for $100. for
Frank's expenses. Let me know his
address - and also let him stay at
school for this winter. I am sure
it will do him much benefit and
be quite a rest to him after the long
work he has had in the field assisting
you. My love to all the family who
I hope are in good health. I send you
a few days since an Army Register
to study while you are confined to
your bed. It is about the only reading
matter I have. I will write you in two
or three days if much of importance happens.
Your Aff Brother
[illegible]

 GR

St Louis Mo Nov 23-1861

Dear Brother

I received letter from home yesterday- and heard you was to be here on 25 or near that time, but have since seen a dispatch from you to your merchants in this place and guage it may be some time before you arrive- write here at Jefferson B'ks- and state the possible time you will be here- I can run up to this city often and wish much to see you.

Dont know how long I shall remain here but sincerely hope a short time. I have seen hell for the last three or four months- and am now very anxious to see the end of this disgusting scrape we are now in - Dont say a word to any one what I write- Give to Frank and say I want to see him very much - We have now 1 senior officer in Washington doing all he can to make an exchange and do all of us hope that he may be successful and that we may once more be on our old footing. Write me before your arrival as I may not miss you - as I often come to town, and we may not meet if you do not inform me beforehand.

Am glad to hear you are doing well that is our merchants here say you are doing well - but then you know they are interested and of course very little weigh can be placed on what they say - No news as to the destination of your regiment or any of the troops in or near Rolla. Our new Genl. has not yet made any disposition. Owing to our present position we see very few of the officers but if I do hear anything which will be of value to you be sure I will write

Your Aff Bro
George

Have not yet had an opportunity of visiting Decatur but hope to able to go soon

(envelope addressed to Mr. M. J. Ryan, Sutler 35-Ill)

Jefferson Barracks Mo.

Nov. 29- 1861

Sir,

I have the honor to recommend for the consideration of the Hon Sec. of War, 1- Serg't George Horney of Co 'A' 7- Infy, for a commission in the Army of the United States.

Serg't Horney entered the service Oct. 1- 1852- joined his company in Texas Dec. 3- same year, was appointed corporal Aug. 1- 1854. Serg't. Jan 1- 1855, and 1- Sergt 10- June 1856, which position he has since occupied,

He has seen much service in the field, in Texas- was on the escort to Mexican Boundary Commission from Sept. 10- 1854 to Nov. 3- 1855, served in Utah from '58 to '60- in New Mexico on the Navajoe Expedition during the winter of '60 & '61-

Serg't Horney is a non-commissioned officer of zeal and intelligence, his education fair- soldierly bearing very fine- his moral character and associations are those of a gentleman, and he has no social ties which could impair his efficiency as an officer or embarrass his social intercourse in the Regiments.

I am. Sir, Very Respectfully
Your Obdt. Serv't
G. Ryan
1- Lt. 7' Infy:
Co 'A'

To;
The Adj't Genl.
U.S. Army
Washington
D.C

Washington D. C. May 26 1862

Sir,

On the 27- July- last- I was surrendered to the Texans by Maj: J. Lynde 7' Infy- at Ft. Fillmore New Mexico- A few days after I was, with the remainder of the command paroled- and am now in this situation.

I would respectfully request- if consistent with the rules which you have established- to be exchanged and placed on active service in the Field-

I entered the Army - with the class that graduated at West Point in 1857- and and joined my Regiment - in the spring of 58- then en-route to Utah- In this Territory I served until May 1860- (a portion of the time on field service- scouting- protecting emigration &c)- when I was appointed Quartermaster and- Commissary to the 2-Sub-column of Trooops ordered to New Mexico- While in New Mexico I was on the Navajoe expedition during the severe winter of 60 & 61- and the remainder of the time on important escort duty or scouting- up to within three days of the approach of the Texans - when I arrived at Ft Fillmore

From the date of the surrender to within a few days I have been constantly with my Regiment- doing the duties of Staff- company- and for the past winter commanding Madison Bks-N.Y

I consider it a matter of duty and expediency to give my parole- as owing to the great distance we were from the States- and the few officers present- our services were necessary for discipline and to take the command across the plains.

I am, Sir. Very Respectfully
Your Obdt. Serv't,
George Ryan
1- Lieutenant 7 Infy:

To:
The Hon. Edwin M Stanton
Sec. of War-
Washington
D.C.
(on back of folded letter)
Washington May 26th 1862

Sec. Edwin Stanton

Hon. E. M. Stanton

Sec. of War.

Sir

As Lt. Ryan is the only officer at surrender of Maj. Lynde who has applied to me to aid him in being placed in active service I must respectfully solicit his early exchange.

Peter L. Watts
New Mexico

The Congress of Delegates from the seceding Southern States convened at Montgomery, Alabama, on February 4, 1861. They quickly adopted a provisional Constitution, and in less than a month, devised and approved a permanent Constitution, which was adopted March 11, 1861.

Ft Niagara June 18

My Dear Parents,

Your letter of 8- was received some days since and I have delayed writing in hopes I might hear something favorable about my exchange. To this time I have heard nothing and fear my prospects are poor. There is at present an effort being made in Congress to have a general exchange but I fear nothing will be done. Buckner is in the way and is the one who has kept us from being exchanged for the last six months, I think it very hard that so many should be obliged to suffer for one man- No matter how grave charges are against him- it would certinly be much better to exchange him - if by so doing other prisoner could be released. I did not present Mr Davis letter from the fact I could not get an audience with Mr Stanton. Both the President and Sec. were very busy about the troops under Banks - and it was impossible for me to remain longer - I did not see Mr Deming- and think he was not in Washington while I was there. Genl. Gillmore was there and we traveled as far as Phild. he being on sick leave and unable to go farther - He wished me to stop at West Point but it was not possible for me to do so, I did not stop in Phild. Say to Grove Lawrence when you see him that I did not have time to stop & see Halsey Stevens in N. Y. Mr. Hitchcock & Hall were one day a head of me coming from Washington - I recognized them the moment we met and spoke to them at once-

Every thing is very quiet here - I have been to the Falls once - Having passed them so often I thought I would take a look - and really say that it is a much finer view than I supposed. There are not yet many visitors and I presume there will not be a great many owing to the prostration of business throughout the country - I have made very few acquaintences - among them a Col. McDougal of the vilage on the other side - I was over at church one Sunday - when he came and spoke to me - and asked me to his farm - and to visit him - He is an old 1812 soldier and appeared to be a very fine man - I

think some day I will go and see him - that is
if I visit any about here

I am enjoying most excellent health

Your aff Son

George

Fort Niagara N. Y.

Sept. 25' 1862

My Dear Parents,

Your letter of 21 - just came to hand - and the one of three weeks since in reference to the colonelcy of the Irish Regiment I received and delayed answering to see if I could do anything. I had no chance to get that Regt. nor have I got the Col. of the N. Y. Reg't. of which I spoke to you - When the appointment came to be made I applied for it in case an army officer was to receive but thot if the appointment was a political one of course I did not expect it. It was a political one - as have all the appointments in this State been under the new call-Capt Potter- who was recently in command here has been appointed Col of a N.H Regt his native state - Crilly has gone to Ft Ontario, leaving me the only officer at this post. our men are not yet exchanged- of course I am daily expecting they will - and as soon as they are off we will be

I am glad Mat has at last been able to get home and only wish I had been there. I am also glad he has invested some of his money in land- for sure if he does meet with reverses- the land will be something he can fall back on- and from the low price at which he bought in comparison to what it was formerly worth- I should judge it would never be much lower. The col. Ryan of whom you speak is not me- at all events if it is I know nothing about it. Ryan you know is becoming quite a common name in this country- and the one you speak of is probably some citizen at the place where the Reg't. is raised

It is extremely annoying after being exchanged to have to remain inactive at a small post like this- in these stirring times. I don't see though how I can help myself- and must therefore wait patiently for something to turn up- My love to all the family who I hope are in good health,

Your aff Son
George

Genl McClellan's Hdqtr

Pleasant Valley, Va.

Oct. 17-1862

My Dear Parents,

At last our men have been exchanged and we are once more fairly in the field. We are at present with the Genl. but wheather we are to form a portion of his body guard over [unreadable] I don't know, Will probably find out in few days- and for the present direct your letters here. I met Genl. Paul, in Frederick city where I was stopped at the time of Stuarts raid- and he told me I was recommended for a col. of Vols- Hope I may get it. Don't like this inactive life. Think I must have some friends in Washington as I have been informed that during last spring- while I was a Lieut. I was strongly urged for a Captaincy in commissary Dept. and for a while it was supposed I would receive it. If I had received it then it would have been of great advantage to me- I am now however a Capt. in the line and of course do not care about. What I want is a Col. of Vols- I expect it will come in due time. I have not been around much- and of course don't know what is going on- The country however is one mass of troops, every hill and valley is covered, for miles and miles- I wish you would tell Mat to write, if he has any way of sending- his letters. I should judge from the newspapers- that he will again be able to follow his regt. to the field - How did Frank come out with the stores that were left behind at Nashville - My love to all- who I hope are well- I will write if we change our position

Your aff Son

George

Genl. Head Quarters
Army of the Potomac
via Baltimore

Headquarters 2nd Div., (Regular) 5th Corps, A. P.,

Camp near Warrenton Va July 29, 1863

Sir-

I have the honor to request that you will inform me wheather you are or are not the author of a letter which appears in the United Presbyterian of July 16- published at Pittsburg Pa- signed J. W. McFarland.

I am Sir Very Resply-
Your Obt Serv't
George Ryan
Capt. 7' Infy-

To:
Rev. J. W. McFarland
Penna [unreadable]

Camp near Fayetteville

July 29th / 63

Capt Ryan

Sir

I have the honor to acknowledge the reception of your note of this evening. I am the author of that letter. I presume you take exception to a remark made in reference to the regulars [at Gettysburg], I have since writing that letter been told that those I saw running were not the regulars, but I had been told that they were & believed it when I wrote, I may have been mistaken

Very respectfully
J. W. McFarland

Envelope addressed to: Assist. Adjt. General of Genl. Sykes 5th Army Corps, via Washington City. Army of Potomac

Camp near Warrenton Va

July 30- 1863.

Sir-

Your note of the 29- inst has been received and as you presume the objectional part was where you spoke of the Regulars

It is much to be regretted that reflections of so serious a character should be made lightly in regard to brother soldier- and as you say that you are now informed and convinced that your remarks were unjust and untrue- I feel confident one occupying your position- representing the principles of truth, justice and charity- will take especial pains to correct at once by an other letter to the same paper the erroneous and false statements contained in your former communication.

When the second letter is published setting forth the truth, I will be much pleased to have you send me a copy- and will then feel satisfied that you erred at first through missapprehension and thoughtlessness

I am Sir Very Respectfully
Your Obdt Serv't.
Geo. Ryan
Capt. 7' Infy
2-Div. 5 Corps Staff

To:
Rev. J. M. McFarland
Penna R. C. 3-Div. 5 Corps-

Camp near Warrenton Junction,

November 4 1863

My Dear Brother-

Your interesting letter reached me while we lay the other side of Culpepper Court House and just before we fell back to Centerville - We had quite a hard march - up night and day - particularly hard on this Reg't - as I had a few days before received a large number of Drafted men from New York- I had quite a rough time looking after them but and find myself quite well off probably have had as good success as any Regtl. commander who have these men- I am still working like a Beaver with them and hope they will do well - when the day of Battle comes- None of them were under fire on the late march- Genl Lee followed us with 66000 men and from the reports of officers captured there is no doubt but the Reb- movement was a complete failure- They did however do much injury to the Rail Road and of course we are delayed some in our present movements. I am so closely confined to my Reg't- as to be unable to hear much news- I presume however that the campaign will not end by our remaining in our present position- am progressig quite well with this Reg't- and I think giving very good satisfaction- I have made quite a number of promotions to officers from the ranks- and think I shall follow the rule - provided good men are to be found- The Gov- of N.Y.- thus far agrees to all my recommendations.

The Reg't is quite large with the addition of the Drafted men- one of the largest in this army- I am now having a Zouave uniform made and as soon as we go into winter will put it on and send you some photographs of the men- In my last letter I promised I think to send you some photographs of myself and forgot to do so- or rather just as I closed the letter we were ordered to move- and in the hurry I packed my valise- and did not keep any out- The letter I put in my pocket and under the impression that we were going into Battle I sent it by first opportunity. We are having very cold weather- and it tells on the men very much- more on those just joined- I have had three deaths within a week- and many more have been sent sick to Washington. I however enjoy most excellent health and trust the same good fortune may continue- Love to all- who I hope are in good health- Hope I may be able to be with you all about Christmas if the circumstances are such that I can get away.

Your aff Brother

George

Camp near Kelly's Ford Va-

November 23, 1863

My Dear Mat,

I am much grieved to hear of your accident- and feel very sorry for the pain I know you must experience- I can imagine nothing more painful than a broken bone- Do you remember the time I fell out of a wagon and hurt my arm when a very small fellow- how I yelled- and made a big fuss- but what was that to a broken leg- I do sincerely hope that by this time you are much better. of course you saw by the papers the account of the little fight we had when crossing this river- We completely surprised the Rebels- now occupy the huts they had built for the winter- In their camps we found orders to them stating that they could make themselves comfortable as they would probably remain here during the winter- We also found many letters which had been written to the soldiers from their friends at home in which they spoke most discourageligly of their prospects one letter from N.C said there as soon as the peace meetings could once get started the people would rush forward and join them with much more eagerness then they had ever gone into secess- I can hardly say what we are going to do- but this much I am quite sure as far as this army is able there will most certainly be a Battle- Genl- Meade is without doubt very anxious to give one- and if Genl Lee will accept I think the probabilities are that in a few days I can give you some news- One thing however looks very bad that is the weather- We have much rain- the roads are terrible and there is much sickness- probably we would lose in the end less men by going into one big fight- then making good quarters for the winter- than by laying around in the mud and water as at present-

Enclosed please find Draft for $100- for Frank's expenses- let me know his address- and also let him stay at school for this winter- I am sure it will do him much benefit and is quite a rest to him after the long work he has had in the field assisting you- My love to all the family who I hope are in good health- I send you a few days since an Army Register to study while you are confined to your bed- It is about the only reading matter I have- I will write you in two or three days if much of importance happens

Your aff Brother
Geo Ryan

NOTES

 GR

Mjr. Isaiah F. Force

The 140th New York drew a good number of men from the Irish and German immigrant population of Rochester

Edward M. Doran

John T. White

camp Warrenton Junction

Va Jan 11 1863 (should read 1864)

My Dear Mat

As this is about the third or fourth attempt I have made to write you I hope it will not be a failure- I had a most miserable trip from home to this place- missed conections on the cars- spent all Christmas travelling in trains loaded down with thousands on the way to their homes to spend the holidays - and made one feel sorry I could not have enjoyed the same in Decatur. Did not see Lt. Grey in Baltimore as the train arrived very late Friday night and next morning I left for Washington, owing to my Reg't having moved I did not reach camp until the 30- I found my new Lt. Col- commanding the Reg't- and from I can see I presume his appointment gives very good satisfaction- We have had very disagreeable cold weather and hard duty to perform- the men only two nights in bed- It has been exceedingly cold- more severe than it was in Decatur while I was there. To day it is a little pleasanter and I am hard at work trying to train my Reg't. I am also at present in command of the Brigade oweing to the temporary absence of some higher officers- I found on my return your letter in answer to mine forwarding you the $100 which should have arrived here a long time before I went home- I also received a letter from Washington at the same time which had been written Oct 2. You can readily see how our mail communications are.

While in Washington I met Dr Bill Welch of Norfolk looking as natural as life with his hands behind his back- head down as if thinking- He did not recognize me - but on making myself known he said ah yes- yes- why George I am very glad to see you- I had quite a talk with him- some four or five hours- he seemed to take much interest in Father and all the family- said that in Norfolk you were reported to be very wealthy- The Dr. was after some gun contract- Have not heard how he succeeded- I spoke to him about Father being re-appointed Postmaster- but as I dont know much about politics I told him I did not know what steps Father had taken- but I hoped he would be re-appointed. He said that probably Father would prefer to get it through Ill- influence but that if he desired Conn- and would write him- that nothing would give him more pleasure than to assist him- he also said that he was acquainted with both of the Conn- Senators. He told me that Ellen Ryan was quite sick and would probably never be any better. My love to all the family who I hope are in good health, I was a good deal under the weather for some days after my return but at present am feeling quite well & hope in a day or so to be as strong as ever- Think you will have some difficulty to read this. Please present my respects to Miss Ewing

Your Aff Bro

George

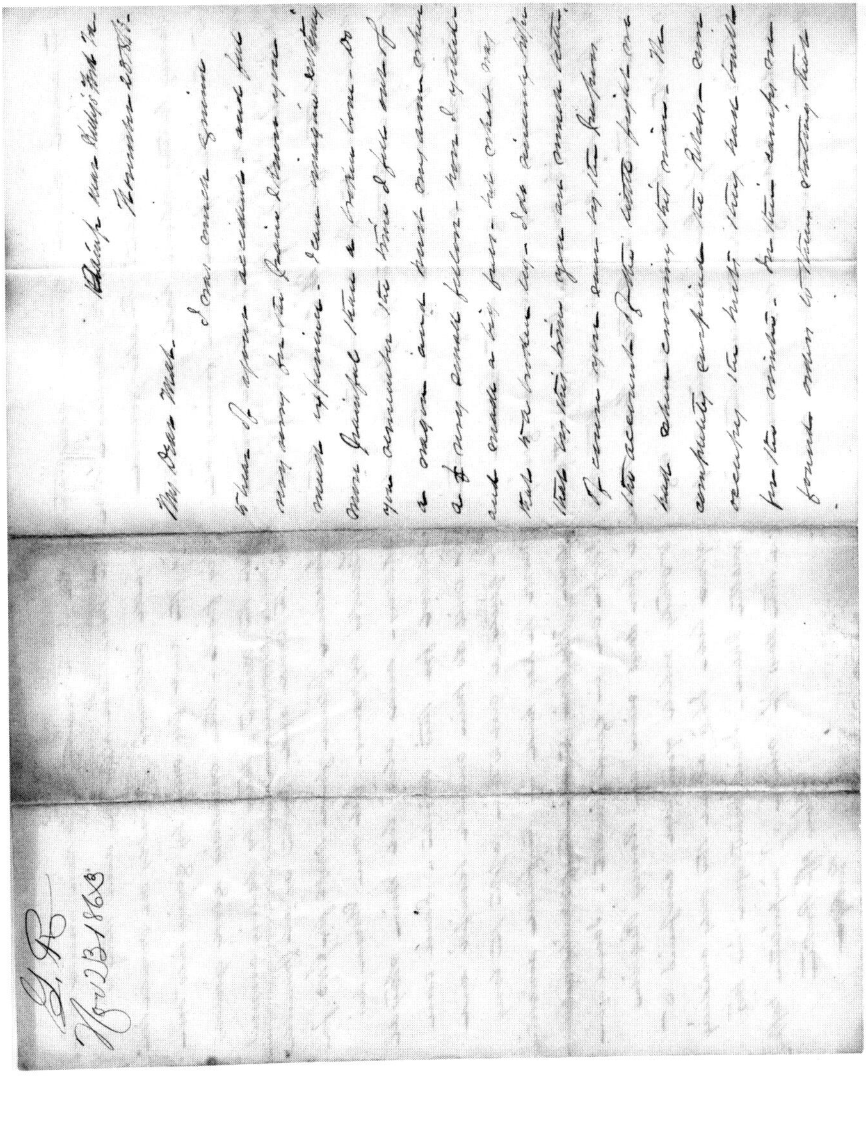

G. R.
Nov 23 1863

Camp near Kelly's Ford Va
November 20/63.

My Dear Mat.

I was much grieved to hear of your accident. and feel very sorry for the pain I know you must experience. I can imagine nothing more painful than a broken bone. Do you remember the time I fell out of a wagon and hurt my arm when a very small fellow. how I yelled and made a big fuss. but what was that to a broken leg. I do sincerely hope that by this time you are much better. Of course you saw by the papers the account of the little fight we had when crossing this river. We completely surprised the Rebels. and occupy the huts they had built for the winter. In their camps we found orders to them stating that

These photos of George's brothers were found in the Ryan family album. Only Bernard "a.k.a. Benny" (top right) can be identified by name

He was a soldier through and through, imbued with the manners and traditions of the regular army to which he belonged... He seemed to have no home or domestic attractions to divert his thoughts from the single purpose of improving the discipline and efficiency of his command. To this work he applied himself with an industry which never flagged. He was a properly modest man and I never heard him allude to his own motives; but his life made plain to every one about him that soldierly duty was his inward watchword, and that a devotion of all his energies to the work involved in the improvement of his command excluded all selfish considerations or ambitions.

— Porter Farley
from "The 140th N.Y. Vols. Wilderness, May 5th, 1864." (delivered on May 9, 1898 to the Rochester Historical Society; published in booklet form, date unknown)

Hd Qr 140th N. Y. Vols-
3rd Brig- 2nd Div- 5th Corps
January 28 - 1864

Received this day of Mr William N. Bancroft Agent- of- G. A. Buel- ($100.#) one hundred Dollars. which I promise to return to him on the receipt of satisfactory evidence that Mr Buel did express to the Mother of E. J. Warner of Co. A. 140th N. Y. V. the sum of $100.# one hundred dolls. In the event of such evidence not being shown- I am to hand the. one hundred $100.# dolls. over to Pvt E. J. Warner- of Co A. 140 N. Y V

Geo Ryan
Col 140th N. Y Vols

Witness,
James R. Campbell
1st Lt and Adjt 140th N. Y. V.

May 7, 1864

My dear John

I have again to be thankful for having passed through a severe engagement without injury - I telegraphed Father to this effect yesterday. My Regiment although repulsed behaved very handsomely. all I could have asked- and just as I expected. My loss is quite heavy ten officers out of twenty one - and (259) two hundred and fifty nine enlisted men out of 488 muskets. I am quite tired and fatigued and must ask you to wait a few days for a full letter You of course know that it is now to early to say much about the Battle

Your Aff Bro

George

Nashville Tenn

May, 16th 1864

Dear Parents

Fathers Dispatch of the 14th instant came to hand a few minutes ago - It would be useless for me to speak of its effect upon Frank & myself - and how much it is to you I can imagine - May God in is mercy, give you the courage & fortitude to bear it as lightly as possible- I have been afraid to look at a paper since the fighting began but not having seen his name among the casulties - I hoped he was safe - the news by me has been partly anticipated but it is hard to believe that such is the case-

My first impulse after receiving the Dispatch was to start immediately for Washington but after thinking the matter over Frank has made me think he can do as well as myself & that it would not be right for me to leave all circumstances considered & I have concluded to let him go - I have every confidence that he will do all that is wanted, and as it may take some days to make the trip & a strong probability that he might be delayed some days there I have concluded to let him go - he will explain to you the situation of affairs here & I hope you will think I have done as I should in the case - Were I to consult my own feelings on the subject I would go immediately regardless of all else but Frank insists on my remaining here & I think you will agree with him- he will explain affairs to you-

He goes to Decatur first to get what information you can give him - I have instructed him to do so that as soon as he gets to Washington if he is unable to accomplish his mission to notify me immediately & I will go there at once - or if on consultation with you you think I had better go - he can telegraph me & I will meet him in Louisville & go on & he can come back-

My Love to all-
Your aff son
Matt

Decatur, Ill, May 18th, 1864

My Dear Son Matt,

Frank arrived here at daylight this morning, and as soon as the telegraph office was open I telegraphed to you. He started on the noon train for Washington. I have received no official account of George's death from any quarters. The only account having reached us being in the Chicago Tribune of Saturday, and that copied from the New York Herald. It is from the regular correspondent of that paper. I have closely examined every paper I could find, but in no other one have seen George's name mentioned. The account is pretty straight, but I have yet some hopes that he may not be killed. In order that you may see the letter of the correspondent of the Herald, I have cut it out of the Tribune and send it enclosed, so that you may read it and judge for yourself. God grant that he may yet turn up among the living. I think Frank will be very thorrow in looking him up.

You have done exactly as I wished to have you in sending him, instead of going yourself.

I know exactly your situation; and I feel very confident he will do as well as you, or any one else could. Gen Oglesby got home this morning and I got a letter, for Frank, from him, to Col Townsend, the assistant Adgt General of the United States. I also gave him a letter to Capt Gray, another to Capt Turnbull and one to Mr Welles, requesting, each to give Frank all the assistance in their power to enable him to find out Georges remains, if killed, and enable him to get them home.

You will perceive by the hand writing, that I am not now very fit for writing letters; but I felt as if I ought to write you this and let you know all I do, in relation to the whole matter, and also what I have done. If George should be dead, I want his remains brout home to us, no matter what the expense, and this I told Frank, and in which he fully agrees with me. I gave him Three Hundred Dollars, besides what he took with him from Nashville, and this, - , if money can get George's remains home, ought to do them.

Your Mother, myself and the whole of us are doing the best we can to bear up under this affliction and hope and pray you will do so too. The journey of life is not beset with roses, and when trials come they must be borne, with fortitude and resignation.

May God bless preserve and protect you, and all the surviving members of the family, is devout and heartfelt prayer of your ever

Affectionate Father
John Ryan

Washington May 20th

Dear Parents

I arrived here to day and met several of Georges Rgt and gentlemen from Rochester who gave me the particulars of Georges death. The account in the Herald was true. I telegraphed immediately both to you and Matt.

I shall go to the front tomorrow and try and get his body-

They say he was decently buried and a head board placed over the grave - They all say that they will do every thing in their power to help me- He was shot in the left shoulder the ball passing through his body and coming out under his right arm. It happened about 8 O'clock in the morning and he lived until evening-

His friends say that they would have telegraphed to you but did not know where you lived only knew it was some where in Ill.

I shall write you as often as possible but if you do not hear from me for several days do not be suprised.

When you write direct your letters care of the Metripolitan Hotel.

I remain as ever your most affectionat Son

Frank

On May 5, 1864, George led 529 New Yorkers across Saunders Field. Savage hand-to-hand fighting ensued and George's troops fell back. "The regiment melted away like snow. Men disappeared as if the earth had swallowed them," said Captain Porter Farley. That evening, George fell into the arms of a fellow officer and tears streamed down his face.

NOTES

86

Gen. Ulysses S. Grant

Washington May 20th

My Dear Matt

I arrived here to day and met several of the 140th, N.Y. and others who gave me the particulars of George's death.

He was shot Friday morning about 8 O'clock but lived until evening - The ball struck him in the left shoulder and came out under his right arm- At the time he was leading his Regt. charging the rebels earthworks. They say that he was buried decently and a head board place over his grave

I shall go front to morrow and do every thing that lays in my power to procure his body.

I know that you must have had a hard time trying to get along since I left. When I arrived home things looked sorrowful enough. Mother feels the loss of poor George to an extent that I can not describe and so with the whole of the family.

Father was perfectly satisfied at your staying in Nashville and so was Mother. both thought that we done the best under the circumstances.

I telegraphed to you upon my arrival here and also to Father. I will write you when opportunity offers but if you should not hear from me for a few days do not be suprised.

I hope and trust that you are in good health and getting along as well as possible under the circumstances- If I get his body will telegraph to you when I start

Hoping that this will find you in good health

I remain as ever Your Affectionate Brother

Frank

-If you write direct letter care of Metripolitan Hotel

Quarter-Master's Office,
No 18 State Street
New York, May 24th, 1864

My Dear Sir

I hardly know how to address you concerning the death of your Son, my most intimate friend in the army. He died a Soldiers death at the head of his Regiment. George and myself were at West Point together and afterwards served in the Same Regiment in New Mexico and other places. I feel his death very Keenly, as do all his friends, and no one had more. His genial disposition, and Soldierly qualities won him the love and esteem of all with whom he was associated.

It is the desire of his Army friends that he should be buried at West Point, if this meets your approval I will be most happy to take charge of everything Connected with the removal of his remains to that place.

He is now buried outside of our lines, but I think in a few days our forces will be sufficiently advanced to enable me to Send Some one there.

I have this day written to an influential friend at General Meades Head Quarters requesting him to Send an ambulance out under a Flag of Truce. Whatever I learn on the subject I will Communicate to you

Very truly
F. J. Crilly
Capt Asst 2.M.
U. S. Army

The footlocker that carried George's effects home from war has the word "Crilly" stenciled on it. Apparently, Francis J. Crilly (above) used his own footlocker to send George's things home. For years, all of the Ryan papers were kept in the old footlocker.

Norfolk Conn May 31 1864

My Dear Brother John

The Document you will receive by the same Mail as this letter came to the Post Office here by yesterdays evening Mail. on being handed to me I perceeved at once it related to George. and as I seen some week or two ago. an acct. of his being. Killed. in the New York Herald at one of the Battles in the Neighbourhood of Spottssylvania Court House. and a few day, after. an account in the same paper stating the loss of the 140th. New York Regement including 10 Line Officers and that. None of the field offecers was injured, I had Strong hopes in in the hurry and confusion of the time was incorrect and had strong hopes he was still safe, from that time. I could hear nothing about him, though I read over every paper I seen hoping to see something about him but did not. till. the document refered to came to hand. On Receipt of it my Anxiety to Know the certaintity of his Fate. overcame my better judgment. and I opened it; between hope & fear, and found my. worst fears realized,

May the Almighty God who gives us all things and who in his Providence takes them away again. when he see fit grant. and. Wife and family Grace and fortitude to bear your affection as becoming a Good Christians.. and grant Poor. George. eternal rest and Peace. in the Kindom of his Heavenly Father which is the Prayer of your Brother

Mathew Ryan

P. S. At another time I will write more at length I do not feel like it now

Your. M, R,

Washington. May 31st

Dear Parents

I got back from the army last night with some of George's effects but was unable to get his body- Where he was buried it is outside our lines and also the enemys main lines but infested by the worst Kind of Gurrillas- Gen Ayres told me that as soon as possible we can get his remains- I have got what he had about his person- his horse run into the enemys lines. I have his mule here and would send it home if possible but as it would be very expensive and I would have to go with it myself I will sell it here I think - I found some of his baggage here at this hotel. Some of his things I could not get very well at present but they will be sent to you. I have concluded to stop a day or two in Philadelphia on my way back- I Know that you will be gratified to Know that the Chaplin of the 9th Mass was with George before he died and administered to him the last rites of the Church.

I received your letter- Will get home as soon as possile

Hoping to see you soon I remain as ever

Your Aff Son
Frank

Gen. Romeyn B. Ayres

Nashville Tenn

June 6th 1864

Dear Father

I received a letter from Frank last evening saying that he was unable to procure George's remains that he should leave for Decatur soon & for me to write him there -- if he has reached home & can be of any use to you for a few days let him remain with you for a while - I can get along here - I am very sorry that he was unsuccesfull but think in a few weeks we may be able to get to the place where he was buried - I have no doubt but that he did all that anyone could do under the circumstances.

I send this by Capt. Peake who goes home in the morning - he can tell you what I am doing - but of course you will take his statements with some allowance. I am not doing anything like as well as he supposes, although I am doing very well. I have got my Permit increased, so that I am now allowed to ship $20,000 per month.

Your aff son Mat

Rochester N. Y.

June 11- 1864

John Ryan Esq

Sir

By request of F. K Ryan Esq. brother of our late beloved Colonel I have the pleasure to forward to you the personal effects of the latter at present in my posession. When I again join the Regiment (after my wound is healed) I Shall try and forward the remainder of his effects now with the Regiment. His Horse was captured by the Enemy. Any information concerning the Colonel you desire will on application be promptly and willingly furnished by

Yours truly
James R. Campbell
1Lt and adj't 140th N. Y. S. V

Present address
Rochester N. Y.

Major James R. Campbell

NOTES

GR

GEORGE RYAN

NOTES

GR

1793. .(Born Mas.)......**GEORGE RYAN**.........(Ap'd Ct.)..**34**

Military History. — Cadet at the Military Academy, July 1, 1853, to July 1, 1857, when he was graduated and promoted in the Army to

BVT. SECOND LIEUT. OF INFANTRY, JULY 1, 1857.

Served: in garrison at Newport Barracks, Ky., 1857–58; on frontier

(SECOND LIEUT., 1ST INFANTRY, OCT. 31, 1857)

duty on Utah Expedition, 1858–60, being engaged in the Capture of Utah

(TRANSFERRED TO 7TH INFANTRY, JUNE 24, 1858)

Indians, Camp, near Box Elder, Aug. 14, 1859, — March to New Mexico, 1860, — Ft. Defiance, N. M., 1860, — Navajo Expedition, 1860–61, —

(FIRST LIEUT., 7TH INFANTRY, APR. 22, 1861)

convoying trains, 1861, — and was captured July 27, 1861, at San Augustine Springs, N. M., by Texas Insurgents, and not exchanged till Aug. 27, 1862.

Served during the Rebellion of the Seceding States, 1861–64: as Adjutant, 7th Infantry, Sep. 1, 1861, to July 9, 1862; in garrison at Jefferson

(CAPTAIN, 7TH INFANTRY, JULY 9, 1862)

Barracks, Mo., and on the Niagara frontier, Nov., 1861, to Oct., 1862; in the Maryland Campaign (Army of the Potomac), Oct.–Nov., 1862, being engaged in a Skirmish at Snicker's Gap, Nov. 3, 1862, on the March to Falmouth, Va.; as Acting Asst. Adjutant-General, 2d Division, 5th Army Corps, Dec. 1, 1862, to Aug. 29, 1863; in the Rappahannock Campaign (Army of the Potomac), Dec., 1862, to June, 1863, being engaged in the Battle of Fredericksburg, Dec. 13, 1862, — and Battle of Chancellorsville, May 2–4, 1863; in the Pennsylvania Campaign (Army of the Potomac), June–July, 1863, being engaged in the Battle of Gettysburg, July 1–3, 1863, — and Pursuit of the enemy to Warrenton, Va., July, 1863; in the Rapidan Campaign (Army of the Potomac), Oct. to Dec.,

(COLONEL, 140TH NEW YORK VOLUNTEERS, AUG. 29, 1863)

1863, being engaged in the Combat of Rappahannock Station, Nov. 7, 1863, — and Actions on the Rapidan and Mine Run, Nov. 24 to Dec. 1, 1863; and in the Richmond Campaign (Army of the Potomac), May, 1864, being engaged in the Battle of the Wilderness, May 5–6, 1864, — and Battle of Spottsylvania, May 8, 1864.

KILLED, MAY 8, 1864, AT THE BATTLE OF SPOTTSYLVANIA, VA.: AGED 28.

They each have stories to tell. The crosses at Normandy. The markers at Punchbowl. The tombs at Arlington. The fallen heroes who rest in places unknown.

From our founding Revolution to today's Global War on Terrorism, nearly one million men and women in the Armed Forces have sacrificed their lives while defending America in time of war.

One hundred and fifty years ago, the Civil War came to an end. Our young country was tested, but remained United. The rememberance to those who fought and died in that war are why we are here, in Norfolk, on Memorial Day. Today I would like to honor one young man who left our town in 1853 to attend West Point at the age of 17. Second Lieutenant George Ryan was assigned to Newport Barracks, Kentucky, the Mormon Campaign, Camp Box Elder, Fort Defiance, Navajo Expedition, and into the Civil War. Through his gallantry he won the admiration of his entire division and was promoted to Colonel of the 140th New York Regiment, by Governor Horatio Seymour in August of 1863. In the Battle of the Wilderness Colonel Ryan was mortally wounded and died on May 8, 1864. Colonel Ryan's final resting place is with the 140th Regiment in Rochester, New York. Colonel Ryan, like so many other defenders of freedom, is forever young.

The numbers of our fallen heroes are not just statistics. They are real people, with real families, who lived in real communities. Let us never forget those who died in the service of their country, for without a witness they just disappear.

Town historian Richard Byrne, Memorial Day 2015, Norfolk, Ct.

AFTERWORD

I wanted to publish these letters for two reasons. First, George was a bachelor when he was killed at 28 and had no children of his own to tell his story. It was left to Mathew and Nancy to keep George alive in memory, and they did.

They passed stories of George down, including a few not found in these letters. Through the years, the family also carried George's Commission papers, his West Point ring, and the old army footlocker that brought his effects home from war.

But the sands of time have had their way, as they always do, and these days there are few left in our family who know who George was. Moreover, George never lived in Decatur or St. Louis, as the other Ryan kids did. For George, the only home he ever knew was Norfolk, Connecticut.

And that brings me to the second reason I wanted to publish this book. Recently, I learned that the picturesque village of Norfolk (pop. 1600) has a town monument dedicated to the memory of the soldiers from this town who died for their country in the war of the rebellion.

Soldiers' Monument stands tall on the Village Green just a pistol shot from the old Puritan meeting house. To my dismay, George's name is not on it. The monument lists 35 names, including a few officers, but no Col. George Ryan. So I decided to publish this book to secure George in memory.

Susan Betts

SALUTE

My thanks to Suzanne Christoff and Casey Madrick of the U.S. Military Academy, who supplied vital information about George and the Academy along with great photos. Marlana Cook, Curator of Art at the West Point Museum, also provided valuable information.

Archivists Shane MacDonald and Paul Kelly of The Catholic University sent along hundreds of letter transcriptions, as well as the beautiful hi-res images found in this book.

Publishers Wayne and Dianne Dementi created the clean layout and design of the book. Wayne, I will miss our weekly editing sessions. Your good humor and easygoing style made this project a joy from start to finish.

Sarah Lapallo Beck of Inkwell Book Co. created our great dust jacket and brought the joys of youth to our book team. Sarah's future is as bright as the coming day!

A heartfelt thanks also to author and Civil War historian Brian Bennett. Brian's work over the years has kept George ever-alive in memory.

This book would not have been possible without the work of Cousin Jerry Goebel, who spent years transcribing each document in the Ryan Family Papers.

And finally, my thanks to Ann Havemeyer, curator of the Norfolk Historical Society. Ann's work has inspired my own, and given my family a new understanding of our own glorious past.

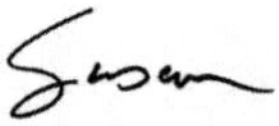

CREDITS

George Ryan, 1857 class album, courtesy of the U. S. Military Academy

"Escaping from the fire in the woods - Wilderness" drawing by Alfred R. Waud, May 7, 1864, courtesy of the Library of Congress

"Position of a Soldier: annihilation of the bowels" pen and ink by James McNeill Whistler, 1852, courtesy of the Library of Congress

Ryan Family Papers doc #329, The American Catholic History Research Center and University Archives, The Catholic University of America, p. 10

Ryan Family Papers doc #310, The American Catholic History Research Center and University Archives, The Catholic University of America, p. 16-17

The Old Academic Building, the U. S. Military Academy, photo by Mathew Brady, p. 42

George Kensel, 1857 class album, courtesy of the U. S. Military Academy, p. 43

Francis Beach, 1857 class album, courtesy of the U. S. Military Academy, p. 43

Official Register of the Officers and Cadets of the U. S. Military Academy, June 1857, p. 49

Ryan Family Papers, The American Catholic History Research Center and University Archives, The Catholic University of America, p. 80

Ryan Family Papers, The American Catholic History Research Center and University Archives, The Catholic University of America, p. 82

Ryan Family Papers, doc # 1005, 1865, The American Catholic History Research and University Archives, The Catholic University of America, p. 83

Ryan Family Papers, doc # 914, 1863, The American Catholic History Research and University Archives, The Catholic University of America, p. 84

Ryan Family Papers, doc #962, The American Catholic History Research Center and University Archives, The Catholic University of America, p. 86

Ryan Family Papers, doc #321, The American Catholic History Research Center and University Archives, The Catholic University of America, p. 87

Ryan Family Papers, doc #802, The American Catholic History Research Center and University Archives, The Catholic University of America, p. 96-97

Ryan Family Papers, doc #934, The American Catholic History Research Center and University Archives, The Catholic University of America, p. 100 and p. 131

The Hon. Edwin M. Stanton, photo by Mathew Brady, courtesy of the Library of Congress, p. 107

Mjr. Isaiah F. Force, courtesy of the New York State Military Museum, p. 127

John Thomas White, courtesy of Anthonny and Beth Zamfoti and the 140th New York Volunteer Infantry Living History Organization, p. 127

Edward Maurice Doran, courtesy of Patrick V. Sweeney and the 140th New York Volunteer Infantry Living History Organization, p. 127

Two soldiers, pencil drawing by Alfred R. Waud, published between 1860-1865, courtesy of the Library of Congress, p. 144

Gen. U. S. Grant at his Cold Harbor, Va., headquarters, June 1864, courtesy of the Library of Congress, photo by Mathew Brady, p. 145

Francis J. Crilly, 1859 class album, courtesy of the U. S. Military Academy, p. 150

Gen. Romeyn B. Ayres, courtesy of the Brady National Photographic Art Gallery, Washigton, D. C. Photo by Mathew Brady, p. 154

Mjr. James Rennick Campbell, photo courtesy of Ronald Erwin and the New York State Military Museum, p. 157

General Cullum's Biographical Register of the Officers and Graduates of the U. S. Military Academy, by George Cullum, Vol. II, p. 160

All of the letters in this book are part of The Ryan Papers collection held at The American Catholic History Research Center and University Archives at The Catholic University of America in Washington, D. C.

The People of the State of New-York,

To

We, reposing especial trust and confidence, as well in you
to do us good and faithful service, by virtue of the power and au
United States and of the State of New-York; Have appo
you the said

NEW-YORK STA

raised under the authority of the President and Congress of
protecting public property," with rank from 186
as you shall from time to time receive from the President of the
other your Superior Officer, according to the Rules and Disci
by the Laws of the United States, in pursuance of the trus

In Testimony Whereof, We ha
Witness, HORATIO SEYMOUR, G
of the same, at our City of Albany, th

Adjutant General.